Water Security for Agriculture:
Atmanirbhar Bharat –
'Water-Smart' & 'Climate-Smart' Action – 2047

Water Security for Agriculture:
Atmanirbhar Bharat –
'Water-Smart' & 'Climate-Smart' Action – 2047

Editors

Prof. Shrikant Kalamkar
Director Professor,
Agro-Economic Research Centre (MOA & FW, GOI), Sardar Patel University,
Vallabh Vidyanagar, Anand, Gujarat, India

Dr. Kirit Shelat
IAS (Rtd.)
Former Principal Secretary, Government of Gujarat &
Executive Chairman, NCCSD, Ahmedabad

Dr. Suresh Acharya
Former Director of Research
SDAU, SK Nagar &R&D Vertical,
AGROCEL Industries Pvt. Ltd. Koday, Mandvi, Kachchh

ALLIED PUBLISHERS PVT. LTD.

Noida • Mumbai • Kolkata • Chennai • Bangalore

ALLIED PUBLISHERS PRIVATE LIMITED

D-5, Sector-2, **Noida**–201 301
Ph. Nos.: 0120-4320295/2542557/4352866
E-mail: delhi.books@alliedpublishers.com

17 Chittaranjan Avenue, **Kolkata**–700072
Ph.: 033-22129618 • E-mail: cal.books@alliedpublishers.com

15 J.N. Heredia Marg, Ballard Estate, **Mumbai**–400001
Ph.: 022-42126969 • E-mail: mumbai.books@alliedpublishers.com

No. 25/10, Commander-in-Chief Road, Ethiraj Lane (Next to Post Office)
Egmore, **Chennai**–600008
Ph.: 044-28223938 • E-mail: chennai.books@alliedpublishers.com

P.B. No. 9932, No. 15, 3rd Floor (Next to Vijaya Bank), 5th Cross, Gandhinagar, Karnataka, **Bangalore**–560009. Ph.: 080-41530285/22386239
E-mail: bngl.books@alliedpublishers.com

Website: www.alliedpublishers.com

ISBN: 978-93-89934-64-9

Citation: Kalamkar, Shrikant S; Shelat, Kirit; and Achayrya, Suresh (2024). (Eds), "Water Security for Agriculture: Atmanirbhar Bharat – 'Water-smart' & 'Climate-smart' Action – 2047", Allied Publishers Pvt. Limited, Noida

Disclaimers: Editors take no responsibility of the statement made by the authors in their papers. It is solely the personal opinion of respective paper author/s.

Published by
Agro-Economic Research Centre, Sardar Patel University, Vallabh Vidyanagar, Anand, Gujarat, India
&
National Council for Climate Change Sustainable Development & Public Leadership, Ahmedabad, India

AERC-Publication Officer and Circulation In-charge
Dr. Deep K. Patel, Res. and Ref. Assistant (Lib)

Published by Sunil Sachdev and printed by Ravi Sachdev
at Allied Publishers Pvt. Ltd., D-5, Sector-2, Noida–201 301

परशोत्तम रूपाला
PARSHOTTAM RUPALA

सत्यमेव जयते

मंत्री
मत्स्यपालन, पशुपालन एवं डेयरी
भारत सरकार
Minister
Fisheries, Animal Husbandry and Dairying
Government of India

D.O. No. 3019.MIN(FAH&D)/20.24....

2 1 MAR 2024

Foreword

Agriculture and allied sectors can be regarded as backbone of Indian economy as it provides livelihood to a large section of the population. It is critical for economic growth as also from the perspective of food security. Agriculture requires large quantities of water for irrigation and of good quality for various production processes. Water is also the most important nutrient to all livestock animals. Thus, water security is pre-condition for crop and livestock production and for food security. Therefore, populous country of Globe having 1.43 billion population need to continuously work on water security for its future development.

World is now facing many challenges; prominent among these are climate change, environmental degradation, sustainable development and food security for growing population. Climate change affects the world's water in complex ways. From unpredictable rainfall patterns to shrinking ice sheets, rising sea levels, floods and droughts, most impacts of climate change come down to water. The effects are already visible and will be catastrophic unless we act now. Through education, innovation and adherence to our climate commitments, and to secure water for agriculture sector keeping in view the Sustainable Development Goals, we can make the necessary changes in agriculture practices/methods to make climate smart/climate re-silent agriculture as well as self-reliant and climate smart farmers. Also, proper use of oceans surrounding our country is also important key to meet the challenges. Ocean provides many resources apart from fishing like sea weed which grows in saline

Delhi Office : Room No. 234, 'B' Wing, Krishi Bhawan, Dr. Rajendra Prasad Road, New Delhi-110 001
Tele.: +91-11-23380780, **Fax:** +91-11-23380783, **E-mail:** rupalaoffice@gmail.com
Gujarat State Camp Office : Plot No. 219, Sector-20, Gandhinagar-382 020, Tele.: 079-23260013
Website : https://dahd.nic.in

water and provide assured income. Sea water can be purified by solar distillation plants used for drinking water and agriculture. Bamboo grows very well on sea shore and also provide excellent income apart from preventing salinity ingress. Also focus need to be given on cultivable waste lands. In fact, due to change in weather pattern in recent past, margin areas on desert of Rajasthan and Gujarat are now emerging as cultivable. This can support livestock-based economy which can provide huge employment and supportive assured income. Solar based distillation plants can be used for drinking water.

Hon'ble Prime Minister Shri Narendra Modiji has announced the 'Atma Nirbhar Bharat Abhiyan' (Self-reliant India Movement) on May 12, 2020. With a motive to transform the agriculture into a sustainable Enterprise, Government of India has taken pro-agriculture landmark initiatives towards the development of Atma Nirbhar Krishi in India. In this context, I am pleased to know that Agro-Economic Research Centre (AERC) of Sardar Patel University, Vallabh Vidyanagar and National Council for Climate Change Sustainable Development (NCCSD)& Public Leadership, Ahmedabad has initiated the academic discussions by organizing the three national seminars/interaction meets by inviting the related policy makers, academicians, stakeholders to contribute their evident based experiences and research for better policy formulation on water Security for agriculture Sector. This volume is outcome of those three interactions meetings. I congratulate AERC of Sardar Patel University, Vallabh Vidyanagar and NCCSD to bring out the edited book on Water Security for Agriculture: Atma Nirbhar Farmers and the editors of book, Dr. Kirit Shelat, Dr. Shrikant Kalamkar and Dr. Suresh Acharya for bringing this valuable book.

(Parshottam Rupala)

Prof. Dr. Niranjan Patel

Vice Chancellor,
Sardar Patel University,
Vallabh Vidyanagar 388120,
Anand, Gujarat

April 19, 2024

Message

Water is a critical input for agricultural production and plays an important role in food security. Irrigated agriculture is, on average, at least twice as productive per unit of land as rainfed agriculture, thereby allowing for more production intensification and crop diversification. As we all are aware, 17 Sustainable Development Goals are targeted to achieve by 2030 in which SDG 13 is focusing on climate action. Climate change is a real and undeniable threat to our entire civilization. Climate change is exacerbating both water scarcity and water-related hazards (such as floods and droughts), as rising temperatures disrupt precipitation patterns and the entire water cycle. Water and climate change are inextricably linked. Through education, innovation and adherence to our climate commitments, we can make the necessary changes to protect the planet. These changes also provide huge opportunities to modernize our infrastructure which will create new jobs and promote greater prosperity across the globe.

Agro-Economic Research Centre (AERC) of our University and National Council for Climate Change Sustainable Development & Public Leadership (NCCSD), Ahmedabad has jointly organized three seminars cum interaction meets on this important topic of livelihood of not only humankind but also whole planet. In fact, AERC since its inception has been working on policy formulation related agriculture and rural development in general and irrigation water in particular. NCCSD is known for its think tank output. I must compliment both these Institutions, viz. AERC and NCCSD for this outcome. I thank policy makers, academicians, stakeholders, bankers, student and all related personal who have contributed in these interaction meets for better policy formulation.

I congratulate the editors of book, Dr. Kirit Shelat, Dr. Shrikant Kalamkar and Dr. Suresh Acharya for bringing this valuable book.

(Prof. Niranjan Patel)

Acknowledgements

This book is the outcome of the presentations and discussion held in three national seminars organized jointly by the National Council for Climate Change Sustainable Development & Public Leadership, Ahmedabad & Agro-Economic Research Centre, Sardar Patel University, Vallabh Vidyanagar, Anand, Gujarat, India.

We have benefited immensely from various scholars, eminent speakers, participants and personalities while conducting the national seminars and preparation of this book. We acknowledge with thanks the support of many personalities who have made this publication possible.

At the outset, we would like to thank **Shri Raghavjibhai Hansrajbhai Patel** (Hon. Minister of Agriculture, Animal Husbandry, Cow-Breeding, Fisheries, Rural Housing and Rural Development, Government of Gujarat); **Dr. V.K. Saraswat** (Hon'ble Member, NITI Aayog, Government of India, New Delhi); **Dr. Meenesh C. Shah** (Chairman, National Dairy Development Board (NDDB), Anand); **Shri Atul Shroff** (Former Managing Director, SFT and Transpek Industry Ltd., Vadodara) for their address at the inaugural session of three national seminar-interaction meets.

We thanks the expert speakers/officers from the Directorate of Agriculture, Government of Gujarat, Gandhinagar; IWMI-Tata Water Policy Research Program, Anand, Gujarat; Gujarat Green Revolution Company Ltd., Fertilizer Nagar Township, Vadodara; Water and Sanitation Management Organisation (WASMO), Ahmedabad; Agricultural Technology Management Agency (ATMA), Gujarat; Farmers from Aatapi Seva Foundation; Aga Khan Rural Support Programme, India; CSR project Coordinator-LOTE, Maharashtra; Anand Agricultural University, Anand; Junagadh Agricultural University, Junagadh; Kamdhenu University, Gandhinagar; Development Support Centre Foundation, Vadodara; Sadguru foundation; Foundation for Ecological Security (FES), Vadod, Anand; all session Chairman, Rapporteurs and Moderator of Session for their participation and support.

We thank **Prof. Niranjan Patel** (Vice-Chancellor) and **Dr. Bhailalbhai Patel** (Registrar) of Sardar Patel University, Vallabh Vidyanagar for their constant encouragement and support for undertaking such academic activity.

We are grateful to the AER Division of Directorate of Economics and Statistics, Ministry of Agriculture and Farmers Welfare, Government of India for its support in organization and publication of this book.

We also thank the staff of AERC and NCCSD for their support in organizing the three interaction meets. Special thanks to **Ms. Nisha Shah** (Chief Executive Officer, NCCSD); **Dr. Kinjal Ahir** (Hon Deputy Director, AERC & HOD, Economics, SPU); **Dr. Hemant Sharma** (AERC, SPU, VVN); and **Ms. Bhavini Aiya** (NCCSD, Ahmedabad) for their support.

Last but not least, we thank all the staff of AERC and NCCSD for their direct and indirect support.

Editors

About the Editors

Dr. Shrikant Kalamkar

Director and Professor
Agro-Economic Research Centre
(Ministry of Agriculture & Farmer Welfare,
Govt. of India)
H.M. Patel Institute of Rural Development,
Sardar Patel University,
Vallabh Vidyanagar 388120,

Dr. S.S. Kalamkar is presently working as Director and Professor in Agro-Economic Research Centre (AERC) of the Ministry of Agriculture and Farmers Welfare, Government of India located at Sardar Patel University, Vallabh Vidyanagar, Anand, Gujarat since July 2012. AERC is established and fully supported by the Directorate of Economics and Statistics, Ministry of Agriculture and Farmers Welfare, Govt. of India. Before joining AERC, he has worked with Gokhale Institute of Politics and Economics (GIPE), Deemed University, Pune (from 2002 to 2012), and ICAR-National Bureau of Soil Survey and Land Use Planning, Nagpur (from 2000 to 2002).

He has made important and useful contributions to the subject of agricultural economics. He has published more than 155 research papers/articles in reputed journals. He has 13 books at his credit. Besides, he has participated in more than 175 national conferences/workshops/seminars. Dr. Kalamkar has completed about 64 research projects/studies, mainly for the Ministry of Agriculture, Govt. of India; the Planning Commission, Govt. of India, Government of Gujarat and NABARD, Mumbai; NDDB, Anand. He has submitted notes on topics of national importance as policy briefs and policy alerts (coordinated and published by CMA, IIM, Ahmedabad) from time to time on to the Hon'ble Prime Minister's Office, the Ministry of Agriculture and Farmers Welfare, GOI & others for consideration in policy formulations.

He is the recipient of Sardar Patel Research Award 2017–18, Shri Dadabhai Navroji Prize (2017–18), SOED-Ludhiana Best Paper Award'; Prof. V.S. Vyas Felicitation award 2014–16 & 2016–18 for the best report completed at AERC. He is also the recipient of the University

Gold Medal and Cash Prize Award for securing the highest CGPA in the subject of Agricultural Economics at M. Sc. (Agri.) degree and three Merit Recognition Prizes.

Besides significant contribution in research, Dr. Kalamkar has also worked as Registrar (Officiating) of the GIPE (Deemed University), Pune from August 2007 to March 2009 as well as Registrar (In-charge) of Sardar Patel University, Vallabh Vidyanagar during March 14, 2015 to April 4, 2015.

Dr. Kirit Shelat (IAS)

Former Principal Secretary,
Government of Gujarat
&
Executive Chairman,
NCCSD, Ahmedabad

Dr. Kirit Shelat is Doctorate in Philosophy with Public Administration. He was awarded a Degree of D. Litt. - Doctorate of Science by Junagadh Agricultural University, Junagadh, Gujarat (India) for his outstanding contribution in promoting 'Climate Smart Agriculture and Building Climate Smart Farmers'. He had long spell of his carrier in Indian Administrative Service. He has hand into introduction in 'New Extension Management- Krishi Mahotsav' approach in Gujarat as Principal Secretary-Agriculture which doubles the income of farmers. He has designed and implemented large scale projects for poor families, farmers and micro entrepreneurs and remote rural areas. He has authored more than 20 books related to agricultural and rural development and related to impact of climate change and ways to meet that challenge at local level-village level. He is Executive Chairman of National Council for Climate Change, Sustainable Development and Public Leadership (NCCSD).

Dr. Kirit Shelat has written biography of Pujya Pramukh Swami Maharaj- "YUG PURUSH, PUJYA MAHANT SWAMI MAHARAJ- a life dedicated to thers". This is published in six languages with nine editions. He has also written biography of Pujya Mahant Swami Maharaj- "Mahan Rushi Mahant Swami Maharaj" – June 2019.

He was member of sub-committee set us by Planning Commission of India on enhancing preparedness for Climate Change and has his hand in introduction of NICRA-National Initiatives for Climate Resilient Agriculture. He was member of expert committee of Government of Gujarat on Economic Revival in Arena Covid Pandemic". Along with Prof Mbuya of FAMU USA, he introduced the concept of "Building Climate Smart Farmers'. His latest publication includes "Atmanirbhar Bharat – "self-reliant and climate smart farmers – Roadmap for Agriculture 2020-30 India" and "Atmanirbhar Farmers – Gujarat – Roadmap 2030".

Dr. Suresh Acharya

Former Director of Research, SDAU, SK Nagar
&
R&D Vertical, AGROCEL Industries Pvt. Ltd.
Koday,
Mandvi, Kachchh

Dr. Suresh Acharya is an expert in Crop Improvement and multi-disciplinary Research Management. After serving in five different State Agricultural Universities of the country. He hund his boots in 2017 as Director of Research, SDAU, SK Nagar, Gujarat. At present, he is working in AGROCEL Industries Pvt. Ltd. Koday, Mandvi, Kachchh, Gujarat.

He has bred 19 versatile varieties (wheat, cotton and pulses), 81 R Lines and 110 A lines in pigeonpea. He guided 12 M.Sc. and 09 PhD Students. He handled 15 diversified projects as Principal Investigator; and organized 16 national level Workshops/Training Courses/Seminars as Organizing Secretary/Nodal officer/Course Coordinator. He has been a member of 17 important committees that also included Extant Varietal Identification Committee, PPV&FRA, New Delhi. He was in the editorial panel of different National Journals and was Editor in Chief of GAU Journal Research.

Consequent upon his expertise, he was conferred ICAR Team Award (1994-96); Sardar Patel Award (2000); Cotton Grower Award (2002); Hari Om Ashram Award (2006); Bharat Jyoti Award (2006); Best Pigeon Pea Team National Award (2008); AND Best Coordinated Project Arid Legume Team (2009).

About the Authors

Dr. Vijay Kumar Saraswat

Member, NITI Aayog, and Former Chancellor, Jawaharlal Nehru University, Delhi

Dr. Vijay Kumar Saraswat is a distinguished scientist with vast experience in defence research—in both basic and applied sciences—spanning several decades. He retired as Secretary, Defence Research and Development Organisation (DRDO), after decades of government service. He holds a PhD from Osmania University and an ME degree from the Indian Institute of Science. Dr. Saraswat has had an illustrious career and is credited with the indigenous development of missiles such as *Prithvi*, *Dhanush*, *Prahaar*, and *Agni-5;* the two-tiered Ballistic Missile Defence system; the initial operational clearance of Light Combat Aircraft *Tejas;* and nuclear submarine *INS Arihant*. He is the recipient of many awards, including the Padma Shri (1998) and Padma Bhushan (2013). He has been conferred an honorary doctorate by more than 25 universities, most recently in 2018 by Jamia Hamdard.

Dr. Meenesh Shah

Chairman & Managing Director, National Dairy Development Board, Anand

Dr. Meenesh Shah has an illustrious career of more than 37 years in Dairy sector and has made immense contribution in the field of Research & Development, Product & Process Development, Project Appraisal & Finance, Cooperative Services, Project Management and Strategic Decision Making. Dr. Shah is also Chairman of MDFVPL, IIL, IDMC, NDDB Dairy Services & NDDB MRIDA, furthering the objectives of NDDB and catering to various value chain of our animal husbandry and dairying sector from animal & human health, dairy machineries, institution building, marketing of milk & milk products to manure management. He is also in the Governing Boards of many other premier institutions of the country. He is Member Secretary of Indian National Committee of International Dairy Federation and also a member of the Standing Committee on Dairy Policies and Economics of IDF. He spearheaded successful organisation of the World's largest Dairy Conference IDF-WDS 2022 in India in September 2022. He had also ensured successful implementation of the World Bank funded Central Sector Scheme 'National Dairy Plan Phase I'.

Dr. Kirit Shelat (IAS)

Former Principal Secretary, Government of Gujarat & Executive Chairman, NCCSD, Ahmedabad

Dr. Kirit Shelat is Doctorate in Philosophy with Public Administration. He was awarded a Degree of D.Litt.- Doctorate of Science by Junagadh Agricultural University, Junagadh, Gujarat (India) for his outstanding contribution in promoting 'Climate Smart Agriculture and Building Climate Smart Farmers'. He had long spell of his carrier in Indian Administrative Service. He has hand into introduction in 'New Extension Management- Krishi Mahotsav' approach in Gujarat as Principal Secretary-Agriculture which doubles the income of farmers. He has designed and implemented large scale projects for poor families, farmers and micro entrepreneurs and remote rural areas. He has authored more than 20 books related to agricultural and rural development and related to impact of climate change and ways to meet that challenge at local leveL-village level. He is Executive Chairman of National Council for Climate Change, Sustainable Development and Public Leadership (NCCSD).

Dr. C.K. Timbadia *Vice Chancellor, Gujarat Natural Farming Science University, Halol, Panchmahal, Gujarat*

Dr. Chandulal Karamshibhai Timbadia is the first Vice Chancellor of new established Gujarat Natural Farming Science University at Halol, Panchmahal, Gujarat by the Government of Gujarat. He has worked as Director of Extension Education, NAU, Navsari; Senior Scientist and Head, Krishi Vigyan Kendra, NAU, Navsari; Project Director of ATMA Project, Tapi. He was nnominated as committee member for preparing syllabus of Natural farming by ICAR, New Delhi and NCERT. He was instrumental as a committee member in preparing syllabus of Natural farming by Government of Gujarat. He has received many awards for his extensive work, few of them are 'Krishi Seva Gujarat Gaurav Award' 2008–09; Best Performing "ATMA District Award" for the year 2009–10; "Krishi Vigyan Kendra Samman"- Mahindra Samriddhi India Agri Award 2014; Best Extension Professional Award 2014 08; Best Krishi Vigyan Kendra Awards (Zonal) from ICAR, New Delhi.

Dr. V.P. Chovatia *Vice Chancellor of Junagadh Agricultural University, Junagadh (Gujarat)*

Dr. V.P. Chovatia is presently the Vice Chancellor of Junagadh Agricultural University, Junagadh (Gujarat). He has served as a Director of Research and Dean PG Studies, Junagadh Agricultural University, Junagadh.

Dr. Suresh Acharya *Former Director of Research, SDAU, SK Nagar & R&D Vertical, AGROCEL Industries Pvt. Ltd. Koday, Mandvi, Kachchh*

Dr. Suresh Acharya is an expert in Crop Improvement and multi- disciplinary Research Management. After serving in five different SAUs of the country. He hund his boots in 2017 as Director of Research, SDAU, SK Nagar, Gujarat. At present, he is working in AGROCEL Industries Pvt. Ltd. Koday, Mandvi, Kachchh, Gujarat. He has bred 19 versatile varieties (wheat, cotton and pulses), 81 R Lines and 110 A lines in pigeonpea. He guided 12 M.Sc. and 09 PhD Students. He handled 15 diversified projects as Principal Investigator; and organized 16 national level Workshops/Training Courses/Seminars as Organizing Secretary/Nodal officer/Course Coordinator. He has been a member of 17 important committees that also included Extant Varietal Identification Committee, PPV & FRA, New Delhi. He was in the editorial panel of different National Journals and was Editor in Chief of GAU Journal Research.

Dr. R. Gopichandran *Professor, NTPC School of Business, Noida, India.*

Dr. R. Gopichandran has designed and tought three courses; Industry Led Sustainability, 'Multilateral environmental agreements & Business Communication and research on policy frameworks for preventive environmental management. He has worked as a Senior Consultant Montreal Protocol & Climate Change, Energy. Efficiency. He has experience in the areas of multilateral environmental agreements especially on compliance assistance; science and technology communication and ecosystems services. Skilled in Nonprofit Organizations, Sustainable Development,

Environmental Issues, and Government engagement. Strong consulting professional with two doctoral degrees in ecology and a degree in law.

Shri Ashwin C. Shroff

Chairman, Excel Industries Ltd., Mumbai, Maharashtra

Mr. Ashwin C. Shroff is a Chairman of Excel Industries Ltd., Mumbai. He is a science graduate and started his career in 1965 as a young trainee and has grown to spearhead Excel Industries Ltd. Along these years of growth, both Excel Industries Ltd. and Mr. Shroff have emerged stronger, helping the growth of Industrial chemicals, Agro chemicals and Environment related businesses. He was the President of Indian Chemical Manufacturers Association (now Indian Chemical Council) during 1996–1998. He has been a leading spokesperson of chemical industry and has represented various issues of the industry to the Government of India. He is connected with various NGOs. He is member of FICCI Environment Committee and FICCI Chemicals Committee. Over and above this he is also holding Directorship in public and private Limited companies such as Excel Industries Ltd. (Chairman & Managing Director); Anshul Specialty Molecules Pvt. Ltd.; Excel Industries (Australia) Pty. Ltd. Australia; Kamaljyot Investment Ltd.; Centre for Bharatiya Management Development; Development & Eco-friendly Enterprises; Federation of Kutch Industries Association.

Dr. Sanjay Deshmukh

Professor of Life Sciences; Former Vice-Chancellor, University of Mumbai, Mumbai

Prof. Sanjay Deshmukh currently holds the esteemed position of Professor of Life Sciences at Mumbai University, and also serves as an

Adjunct Professor at the Institute of Chemical Technology, Mumbai. He is a distinguished academic thought leader in India, renowned for his exceptional leadership qualities, commitment to organizational success, and extensive educational background. As the youngest full-time Vice-Chancellor of the University of Mumbai, Prof. Deshmukh spearheaded its remarkable transformation characterized by transparency and the implementation of global best practices. His focus on skill development and employability aimed to equip students for the professional world. Prof. Deshmukh's academic prowess is reflected in his diverse educational qualifications, encompassing degrees in Science and Law. This multidisciplinary approach fuels his innovative thinking and problem-solving abilities, contributing to his exceptional achievements. He has been bestowed with two honorary DSc Degrees.

Shri Atul G. Shroff

Managing Director at Excel Industries Ltd., Mumbai, Maharashtra, India

Mr. Atul G. Shroff, who had been the Managing Director of the Company from 1981 to November 2017, has grown through the ranks of the Company since its inception in 1965. A people's man, he is known to be found moving on the shop-floor, and having a direct rapport with most of its workforce which constitutes the Transpek family. His vision envisages the setting up of a global network. His keen sense of understanding the chemical processes has resulted in development of several innovative and appropriate improvements in various plants. He is committed to develop Transpek into a world class company, where each employee is a proud partner. He strives to harness the most innovative and appropriate technologies for manufacturing chemicals having complex chemistry with due concern to the environment and safety of all employees and the neighbourhood.

Dr. A. R. Pathak

Former Vice Chancellor, Junagadh Agricultural University, Junagadh

Dr. A.P. Pathak was a Vice-Chancellor of Junagadh Agricultural University (JAU) and also had hold charge of Navsari Agricultural University. Earlier, Dr. Pathak had also served as Director of Research and Dean of Post-Graduate Studies in Anand Agricultural University for 4 years. He has specialization in plant breeding and genetics. He had developed twenty-five new varieties of different crops which include cotton, mungbean, clusterbean, pigeonpea, cowpea, chickpea, rapeseed, mustard, castor varieties and released for commercial purpose.

Dr. P.A. Pandya

Assistant Professor at Junagadh Agricultural University, Junagadh (Guj.), India

Dr. P.A. Pandya is presently Assistant Professor at Junagadh Agricultural University, Junagadh (Guj.), India. He has worked/Working on Drip Irrigation, Agricultural and Meteorological Drought characterization and development of operation agricultural drought monitoring applications using remote sensing and machine learning, Hydrology, Watershed Management, Soil and Water Conservation. Teaching to UG (B.Tech, Agril. Engineering) and PG (M.Tech, Soil and Water Conservation/Irrigation and Drainage Engineering) at JAU, Junagadh, Gujarat, India

Dr. M.L. Gaur — ***Professor (SWC Engg), Deptt of Meteorology and Agril Engg, Anand Agril University, Anand***

Dr. M.L. Gaur is now Professor (SWC Engg), Deptt of Meteorology and Agril Engg, B A College of Agriculture, Anand Agricultural University, Anand. His professional specialization is Soil & Water Engg, Watershed Management, Hydrology, Soil Conservation, Grassland & Forest Hydrology, GIS and Remote Sensing, Water Resources.

Dr. H.D. Rank — ***Professor, Department of Soil & Water Engineering, Junagadh Agricultural University, Junagadh***

Dr. H.D. Rank Currently working as Professor in the department of Soil & Water Engineering in Junagadh Agricultural University. His areas of interest are Water Resources Development & Management.

Prof. Sonal Bhatt — ***Professor in the Department of Economics, Sardar Patel University, Vallabh Vidyanagar***

Prof. (Dr.) Sonal Bhatt is currently working as a Professor in the Department of Economics, Sardar Patel University, Vallabh Vidyanagar. She has various academic accolades including NUFFIC Fellowship from the government of Netherlands (twice), 'Dadabhai Navroji Award for Best Research Paper' by Sardar Patel University (five times), Second Prize by the National Institute of Public Finance and Policy, New Delhi for her research paper; a Gold Medal in M.A. Economics by Sardar Patel University as well as a Doctoral Fellowship from Indian Council of Social Science Research, New Delhi. She has a teaching and research experience of twenty-three years. Her research interests include the economics and management of the natural resources as well as public

	policy and developmental issues. She has published twenty-seven research papers in national and international journals and has completed eight research projects. She has guided five M.Phil. and one Ph.D. student and is currently guiding two Ph.D. scholars.
Shri Karthik Dongre	*Scientist-D and Head of Office, CGWB, Ahmedabad*
Ms. Bharti Bhavsar	*Rural Development Department, Clean Sky Campaign, Self Employed Women's Association (SEWA), Ahmedabad/Anand*
Dr. Kanhaiya Choudhary	*Chief Executive Officer, N.M. Sadguru Water and Development Foundation*
Dr. Mohan Sharma	*Director, Development Support Centre* Dr. Mohan Sharma is a agriculture Engineer and Management professional, he has 18 years of experience in the field of participatory natural resource management and rural livelihoods. He has been an action researcher and trainer on subjects such as irrigation, soil and water management, farmer producer organization and institutional development. He has been providing technical services to several states for capacity building of Govt. officials and Farmer Organizations in the field of PNRM. He has a bachelor's degree in Agriculture Engineering and holds Master of Business Administration (Marketing). He is involved in DSC foundation`s activities as research advisor and lead trainer.
Shri Mandar Prabhune	*Technical Head, Transpek Industries Ltd.*
Dr. Shrikant S. Kalamkar	*Director Professor, Agro-Economic Research Centre, Sardar Patel University, Vallabh Vidyanagar* Dr. S.S. Kalamkar is presently working as

Director and Professor in Agro-Economic Research Centre (AERC) of the Ministry of Agriculture and Farmers Welfare, Government of India located at Sardar Patel University, Vallabh Vidyanagar, Anand, Gujarat. He has made important and useful contributions to the subject of agricultural economics. He has published more than 155 research papers/ articles in reputed journals. He has 13 books at his credit. Besides, he has participated in more than 175 national conferences/workshops/ seminars. Dr. Kalamkar has completed about 64 research projects/studies, mainly for the Ministry of Agriculture, Govt. of India; the Planning Commission, Govt. of India, Government of Gujarat and NABARD, Mumbai; NDDB, Anand. He has submitted notes on topics of national importance as policy briefs and policy alerts (coordinated and published by CMA, IIM, Ahmedabad) from time to time on to the Hon'ble Prime Minister's Office, the Ministry of Agriculture and Farmers Welfare, GOI & others for consideration in policy formulations.

About the Book

Water Security for Agriculture Sector: Summary

by

Shrikant S. Kalamkar, Kirit N. Shelat & Suresh Scharya

Introduction

India is an agriculture economy where land and water are two key natural resources upon which farmers depend for their livelihoods and development. Farmers' development depends upon interactions of these and other resources, institutions, actions and policies and their ultimate outcomes. It would be naive to perceive that all rural poverty problems could be solved through improving the poor's access to water alone through development of irrigated area in rainfed conditions. However, though water is only a single element in the poverty equation, it plays a disproportionately powerful role through its wider impacts on such factors as food and other essential agricultural production. Water is one of the most critical inputs for agriculture. The availability of adequate water for irrigation is a key factor in achieving higher productivity. However, the poor efficiency of conventional irrigation systems has not only reduced the anticipated outcome of investments towards water resource development, but has also resulted in environmental problems like water logging and soil salinity, thereby adversely affecting crop yields.

Water, being a necessity for crop production, is one of the most important natural resources for sustaining human life on earth. However, owing to the presence of large tracts of arid and semi-arid lands, where the surface and sub-surface water resources are highly limited, coupled with the spurt in industrial & domestic consumption of water due to a high rate of population growth, the competition for this limited commodity is increasing day-by-day in the country. Further, the over-exploitation is depleting the existing water resources at critical rates even in areas hitherto known for their having irrigation water in aplenty, resulting in irrigation water becoming both scarce and expensive. Thus, to feed the ever-growing population, the agricultural production needs to be boosted by following better soil-water management techniques that could provide the arid and semi-arid lands better access to irrigation water without actually increasing the stress on available water resources by changing the cultivation practises to mitigate the impact of climate change. Higher temperatures are projected to affect all aspects of the hydrological cycle. More frequent and severe droughts and floods are already apparent, and their impact increases as a growing population becomes more dependent upon a set of atmospheric and hydrological circulations. Climate change will impact the extent and productivity of both irrigated and rainfed agriculture. Rising temperatures will translate into increased crop water demand, so will be demand of increasing Urbanities. Both the livelihoods of rural communities and the food security of a predominantly urban population are therefore at risk from water-related impacts linked primarily to climate variability. The rural poor, who are the most vulnerable, are likely to be disproportionately affected. Various adaptation measures that deal with climate variability and build upon improved land and water management practices have the potential to create resilience to climate change and to enhance water security. In India, distribution of rainfall is highly non-uniform both in terms of time and space. As a result, water is required to be stored and utilized for meeting the demands of different sectors throughout the year. Efficient water management requires sustainable development of the available surface and ground water resources and their optimal utilizations including enhanced use of micro-irrigation system. In order to meet the challenge of Food Security, first, there is a need to understand the need of water security. Therefore, three national interaction meets were organized to deliberate on climate change and its impact on water cycle, water Security, and that in turn on Food Security. The final papers come out form the eminent expert in these meets are briefly summarized here.

Dr. V.K. Saraswat in his article on 'Water Security in Agriculture – Identifying Advanced Technology Confluence' noted that achieving water security in the agricultural sector is a multifaceted challenge that demands a holistic approach. Advanced technologies present a transformative opportunity to address water scarcity issues and enhance agricultural productivity, by embracing the prospects of AI and ML, IoT, digitalisation and circular economy principles, ensuring a more sustainable and water-secure agricultural landscape. As we navigate towards the Sustainable Development Goals (SDGs) for 2030, integration of these technologies into mainstream agricultural practices becomes not just an option but also a necessity for a water-secure and food-sufficient India.

While **Dr. Meenesh Shah** in his article 'Water Security for Agriculture' stated that addressing water security requires a collaborative and multi-stakeholder approach, with the active engagement of Government, civil society, cooperative and private sectors, and farmers. Investing in water security would improve water availability, quality and contribute to the overall socio-economic and sustainable development of the country. He suggested for sustainable use of groundwater through solar energy.

Dr. Kirit N. Shelat & Dr. R. Gopichandran in their article on 'Integrated Water Resource Management at the centre of Climate Resilient Agriculture: Some immediate Local level Considerations & Emerging Trends' highlighted the snapshot of emerging considerations on the water/agriculture/sustainability interface. It points to the robustness of India's approach to sustainable development through a systems approach. The effort due not pretend even to a minuscule extent about anything scholarly. It is a clear eclectic exercise to help beginners in the field of agriculture studies to know about the interrelatedness of SDGs with a special focus on water and agriculture. Sources of important leads of information in this context are presented to help easy access for deeper learning.

Dr. C.K. Timbadia in his article on 'Water-Resilient Agriculture: A Holistic Approach through Natural Farming' explored Natural Farming practice for enhancing water resilience and sustainability in agriculture. The research emphasizes the pivotal role of earthworms, in improving soil moisture retention, infiltration rate, and overall water-holding capacity. Additionally, the application of Jeevamrit in natural farming demonstrates significant increases in soil organic carbon, microbial diversity, and water-holding capacity. Combining zero tillage with cover crops proves effective in enhancing water availability and improving soil structure, while crop diversification emerges as a key strategy for water conservation. The study highlights the positive correlation between soil organic matter and moisture retention, underscoring the importance of these factors in mitigating the drought risks. By prioritizing these natural and regenerative approaches, agriculture can not only adapt to water-related challenges, but also contribute to the broader goals of sustainability and ecological balance

Paper by **Dr. V.P. Chovatia & Dr. H.D. Rank** on 'Status and Issues of Water and Food Securities in Gujarat' highlighted that Gujarat, despite being a surplus state in food production, faces challenges in water and food security due to reliance on rainfed farming, erratic monsoons, and inefficient irrigation. Groundwater withdrawals worsen soil degradation, seawater intrusions. Farmers struggle with volatile market prices, limited technology access, and debt burdens. The water and food securities in the state should be enhanced by investing in water harvesting, water transfer by river linking, efficient irrigation water management and preventing seawater intrusion, adoption of climate-resilient crops, and organic farming. Strengthening market access, fair pricing mechanisms, and social safety nets are crucial for farmer well-being. Infrastructural development should be in fallow/wastelands to retain the per capita availabilities of land.

Dr. A.R. Pathak and Dr. P.A. Pandya in their paper on 'Water Security for Agriculture -Atmnirbhar Farmers' highlighted the importance of water security for making Indian Agriculture and Farmers Atmnirbhar. The article highlighted the critical connection between agriculture, water security, and global concerns regarding water scarcity. In India, water security faces challenges such as over-extraction of groundwater, pollution, and unequal distribution. Agriculture, being the largest consumer of water globally, necessitates innovative solutions. Technological interventions, including Micro Irrigation Systems (MIS), water harvesting, and smart irrigation, can enhance water efficiency. Remote sensing and GIS technologies offer valuable tools for monitoring and managing water resources. Mobile applications aid farmers with real-time information, while alternative cropping systems and biotechnological interventions contribute to water-efficient practices. Soil health management, irrigation water quality, deficit irrigation, and mulching technologies further address sustainable water use. The paper emphasizes the multifaceted nature of water security, especially in the context of agriculture, and advocates for a holistic and collaborative approach involving government, civil society, private sector, and communities to ensure sustainable water management and overall economic and social development. To make more meaningful and efficient contributions to addressing water security and resilience across various sectors and scales, there is a need for interdisciplinary research and the establishment of productive synergies between policymakers and researchers.

Dr. Murari Lal Gaur in his paper on 'Water Scarcity and Security in Agriculture: Challenges and Resolutions in India' explored the dynamics of water security in agriculture in India, examining the challenges faced and proposing sustainable solutions to mitigate these challenges. It also incorporates food for thought on critical issues like major obstacles in agricultural water management, climate change impacts on water security

& scarcities, projected water securities for next few decades, and also some of the strategic adaptations for such scarcities with relevant elaborations on gauged efforts, reasoning and solution matrix. Lastly, role of digital tools and technologies and related inventions & interventions towards smart water management in farming sectors are made part of writeup, to project some of the latest inventions, tools, technologies and potential options to overcome water securities and scarcities in agricultural and allied sectors. A summarised elaboration is offered at last with most updated references of key nature.

Dr. Suresh Acharya in his paper on 'Water Resilient Agriculture' presented complexities of the SW-MEALS (Soil-Water-Microbes-Ecology-Agriculture-Livelihood Security) nexus in water-resilient agriculture requires a comprehensive strategy. Balancing the 'Demand Side' with the 'Supply Side,' weather conditions, and the natural resource base is crucial. The synergy between traditional methods and cutting-edge technologies is indispensable for a sustainable and resilient agricultural future. A participatory approach involving farmers and communities in local water budgeting, with incentives for water savings, is vital for promoting responsible water use. Innovative technologies, including smart irrigation systems, antitranspirants, anti-evaporation tools, and biocapsules, play a significant role in optimizing water management. Achieving water-resilient agriculture amid climate change requires a harmonious blend of traditional wisdom and innovative technologies, emphasizing the need for comprehensive strategies to enhance water resilience.

Paper by **Dr. Sanjay Deshmukh** on 'Enhancing Water Security for Resilient Agriculture in India' highlighted the nexus of water security and agricultural productivity unfolds a realm of opportunities for farmers, acting as a beacon of hope in the face of formidable challenges. By embracing water-smart practices, farmers can not only enhance their resilience to climate variability but also contribute significantly to sustainable water use. The integration of efficient irrigation systems, rainwater harvesting, and soil moisture management techniques presents promising avenues for optimizing water resources.

Moreover, the exploration of alternative crops that are well-suited to local agro-climatic conditions can diversify income streams for farmers. This diversification, coupled with access to markets for water-efficient produce, contributes to economic sustainability for farming communities. In the rain-fed regions of Andhra Pradesh, the adoption of precision agriculture techniques, coupled with the cultivation of climate-resilient crops like pulses and oilseeds, has transformed the agricultural landscape. This paradigm shift not only enhances water-use efficiency but also elevates farmers' incomes.

Dr. Ashwin Shroff in his article on 'Water Security and Novel Solutions for Agriculture' shared that Excel group has been actively engaged with Agriculture, farmers and rural communities. There are several initiatives, many for the first time in India and even globally eg. Chemical pesticides, drip systems. Compost from organic wastes, sewage water treatment.

A paper by **Dr. Sonal Bhatt** is an attempt to understand the implementation of SKY in Anand district of Gujarat; impact of SKY on the withdrawal of groundwater, groundwater table, its price and the area irrigated with the help of ground water by various stake-holders such as SKY sellers, non-SKY sellers (water sellers without SKY connection) and water buyers and to understand the importance of non-economic factors viz-a-viz economic factors in the determination of price of groundwater. The study found that Farmers expressed their dissatisfaction with all the regimes of electricity supply and price implemented till date, for various reasons. However, by far, their satisfaction with flat rate regime was highest in terms of price of power and with JGY, it was highest in terms of service provided. The water rates for irrigation service have been increasing periodically, but have risen very rapidly and almost doubled under the SKY regime. The SKY sellers were aggrieved because they were ill-informed about the financial arrangements of SKY and not properly oriented on how they could become net producers of power instead of net consumers. In course of time, they started learning and few farmers had also started earning net income from selling power under SKY. Majority of respondents did not pay up their electricity bills and had no idea how they would face the cumulative effects of their non-payment. However, they had shifted the burden on their water buyers either increasing the water rates, decreasing the irrigated land, decreasing the number of irrigations or stopping the sale of water altogether. Water buyers resented the fact that SKY sellers had increased the water rates in spite of not actually paying up the electricity bills owed by them. However, they were happy with SKY due to the increased hours of electricity supply and its more convenient timings. The non-SKY sellers did not increase their water rates after SKY. However, they felt left out because they could not sell as much water as the SKY sellers, since they had only 8 hours of electricity supply. Thus, while some aspects of SKY were welcomed, it has been the cause of a stark increase in water rates, which however, did not have a substantial impact on the area irrigated, number of water buyers or the amount of water supplied to them.

Dr. Atul G. Shroff & Shri Mandar Prabhune in their paper on 'Water Security for Agriculture Sector: Reuse & Recycling of Treated Sewage for Agriculture' highlighted that the greatest challenge in the Water and Sanitation sector over the next two decades will be the implementation of 'Low-Cost Sewage Treatment' that will at the same time permit selective REUSE of treated effluents (disinfected & detoxified) for agricultural & industrial purposes.

Dr. Mohan Sharma shared the initiatives of Development Support Centre marching towards its 29 years journey in Promoting Water Security for resilient agriculture. He suggested to realign the programs and policies to create an enabling environment for facilitating water users, practitioners, and donors in achieving water security.

Ms. Bhartiben Bhavsar shared work of SEWA in natural resource management campaign and role of women in water conservation. She shared that the districts most affected by water insecurity include Rajkot, Bhavnagar, Amreli and Jamnagar in the Saurashtra region and Banaskantha, Surendranagar, Patan and Mehsana in north Gujarat. In several villages of Patan and Banskanatha, the ground water level has dropped and content of fluoride in the water has increased alarmingly, posing health hazards to the community.

Mr. Kanhaiya Choudhary shared initiatives of N.M. Sadguru Water and Development Foundation in his article 'Water Security for Agriculture - Atma Nirbhar Farmers'. He concluded that water security at first place has been key to success of all the development efforts in N.M. Sadguru's work area. The organisation gave the prime importance to water security right from beginning which was heartfelt need of the community and the region. Agriculture based livelihood has become remunerative and regenerative with water security.

Dr. Kartik P. Dongre in his article 'Water Security for Agriculture' stated that Groundwater is the backbone of India's agriculture and drinking water security. It is a common-pool resource, used by millions of farmers across the country and remains the only drinking water source in most of India's rural households. He suggested that there is a need for strategies aimed at scientific and sustainable management of the available ground water resources in the country. Such strategies should not only address the issue of increasing the ground water availability, it should also consider the aspects of Regulatory & Protection Measures, as well as Public Awareness regarding various aspects of ground water.

Contents

Water Security in Agriculture—Identifying Advanced Technology Confluence

V.K. Saraswat[1]

Introduction

Agriculture, the backbone of global food production, faces a crucial juncture in water management, presenting both challenges and opportunities. The United Nations predicts a 50% increase in agricultural production by 2050, raising concerns about water scarcity, inefficient irrigation practices and environmental impacts. To sustain a daily diet of 2,800 kilocalories, each person requires 2,000 to 5,000 litres of water per day. Consequently, the United Nations forecasts a 30% increase in global water withdrawals for agriculture by 2050. Currently, 2.4 billion people experience water stress, with nearly 40% of global croplands facing water scarcity. Small-scale farmers, who contribute significantly to global food production, struggle with water challenges exacerbated by climate change.

Indian Water Landscape

According to the Indian Agricultural Statistics Report 2022, the agriculture sector employs over 50% of the Indian workforce, especially in rural areas and contributes 18.8% to the country's GDP. However, India faces a significant challenge with its water resources. Despite having 18% of the world's population, it has only 4% of its water resources. Agriculture in India consumes a staggering 90% of the country's water, far exceeding the global average of 72%. The nation's per capita water availability has declined from 5000 cubic metres (m^3) per annum in 1950 to around 2000 m^3 and it is projected to decline further to 1500 m^3 by 2025, leading to significantly reduced water availability for agriculture.

Present Scenario of Concern

Factors such as over-irrigation, inadequate water distribution, climate change, population growth, soil erosion and pollution place immense

[1]Member, NITI Aayog, GOI, New Delhi.

pressure on agricultural water resources. The challenges are compounded by conventional crops that lack micro-irrigation adoption, neglected storage capacities and poor water infrastructure, all contributing to water scarcity. In addition, farmers in many regions face increasing competition from non-agricultural users due to rising urban population density and water demands from the energy and industry sectors.

Addressing future water challenges in agriculture necessitates a systematic re-evaluation of water management practices and an overall reassessment of water resources management and policies in agriculture. Current institutional policies, whether at the national or international level, are insufficient to effectively manage agricultural water, thus jeopardizing future water and food security.

In the agricultural framework, both supply- and demand-side measures are essential for effective water management, with strategies chosen based on their effectiveness and ease of implementation. Demand-side measures, such as structural and operational changes (e.g., replacing inefficient water pumps, adopting drip irrigation and using laser land levelling), economic incentives (e.g., financial rewards for reducing water waste in irrigation) and providing training and education to farmers on water-saving technologies, can play a crucial role.

On the other hand, supply-side management techniques can be expensive, posing financial barriers to their implementation in developing countries. However, measures such as implementing small farm-sized dams and growing crops based on water requirements and availability are commendable approaches.

Key Interventions

A multitude of interventions offers a comprehensive framework for sustainable water management.

- **Technical Interventions:** Bringing Crop & Livestock closer together with manure management technologies.
- **Social Interventions:** Combining scientific & practical knowledge through education
- **Economic Interventions:** Accounting for cost-effectiveness in solutions.
- **Policy Interventions:** Consider synergies and trade-off between agricultural practices and the environment in spatial planning.

- **Environmental Interventions:** Building on nutrient recycling & ecological resilience.

Best Agricultural Practices for Water Conservation

Drip irrigation, irrigation scheduling, dry farming, composting, mulching, cover cropping, capturing and storing water, drought-resistant crops, rotational grazing, conservation tillage and organic farming are essential practices for water conservation. Additionally, rainwater harvesting, desalination, greywater recycling and aquifer recharge play crucial roles. Equally important for water quality protection are nature-based solutions (NBS), which leverage natural processes to address socio-environmental challenges. These solutions mimic nature to establish a symbiotic relationship between human activities and the environment, thereby enhancing ecosystem health and resilience.

Advanced Technology Approaches

To achieve sustainable water management strategies, leveraging advanced technologies is crucial. A variety of technologies can enhance water efficiency, ensuring that each unit of water used in agriculture produces the maximum possible value.

AI-Driven Precision Agriculture

Integrating AI-driven technologies such as machine learning, computer vision and remote sensing enables farmers to collect and analyse large volumes of data in real-time. These data-driven insights help optimize farming practices, including irrigation, fertilization and pest management. As a result, AI-driven precision agriculture can significantly improve water efficiency and contribute to sustainable farming practices.

Smart Irrigation Systems

These systems utilize various sensors, such as soil moisture sensors and weather data, to monitor the water requirements of crops. Machine learning algorithms then process this data and offer real-time irrigation recommendations. This approach enables farmers to apply water only when and where needed, reducing water waste and improving overall efficiency.

Predictive Analytics and Crop Modelling

AI models can predict crop water needs and growth stages by integrating historical data, weather patterns and crop-specific information. These insights are used to optimize irrigation schedules and adapt to changing environmental conditions.

Satellite and Drone-Based Remote Sensing

Remote sensing technologies, such as satellite imagery and drones (UAVs—unmanned aerial vehicles), are increasingly used to monitor crop health, soil moisture and other environmental parameters. AI algorithms can analyse these large datasets to identify patterns and trends related to water use efficiency. For example, remote sensing can detect areas of over-irrigation or water stress, enabling farmers to make targeted adjustments to their irrigation practices.

Internet of Things (IoT) in Agriculture

IoT applications in agriculture involve integrating smart devices and sensors connected through the internet. In water management, IoT enables real-time monitoring of water infrastructure, automated irrigation systems and data-driven decision-making. This interconnected ecosystem enhances efficiency and reduces water wastage.

Other Emerging Technologies

Nanotechnology for water purification, machine learning algorithms for predictive analytics and the integration of blockchain for transparent water transactions are some of the emerging trends that hold promise for the future. Moreover, continued research and development in the field are giving rise to innovative technologies.

Circular Economy approach

A circular economy in agriculture implies a regenerative system that reduces resource input, waste and environmental degradation. The primary goal is to close the loop on resource use, creating a system where waste is minimized and resources, including water, are recycled and reused.

Start-up Contributions

Start-ups worldwide are playing a pivotal role in driving innovation in agricultural water management. Companies are actively developing cutting-edge solutions, from AI-powered irrigation controllers to satellite-based precision farming tools. Their contributions are fostering a competitive market and accelerating the adoption of advanced technologies. India is prominently positioned in the global start-up heat-map for water management (*Figure 1*). Strategies involving technology and innovation investments, rainwater harvesting, spate irrigation and renewable energy pumps showcase India's commitment to addressing water challenges.

Market Projections

The market for advanced technologies in agricultural water management is experiencing robust growth. According to the latest projections, the precision farming market is estimated to reach around USD 34.01 billion by 2032 with a CAGR of 13.30%. Similarly, the smart agriculture market is expected to grow from USD 16.2 billion in 2022 to USD 30.4 billion in 2030 with a CAGR of 9.4%, in the Asia-Pacific region alone. In India, the smart agriculture market's growth is anticipated to progress at a CAGR of 13.38% during 2022–2028 and is expected to garner a revenue of USD 886.21 million by 2028. The national market is poised for rapid expansion, driven by government initiatives, increasing awareness and the imperative for sustainable farming practices the smart agriculture.

The national market is poised for rapid expansion, driven by government initiatives, increasing awareness and the imperative for sustainable farming practices.

Figure 1: Start-up Global Presence

(*Source:* Startus Insights)

Government Initiatives

The Government of India has initiated targeted schemes to address the pressing needs of the agricultural sector, including the *ATAL Bhujal Yojana*, National Mission for Sustainable Agriculture (NMSA), *Pradhan Mantri Krishi Sinchai Yojana* (PMKSY), Accelerated Irrigation Benefit Programme (AIBP) and Command Area Development & Water Management (CADWM).

These schemes are intended towards improved utilization of water resources, water distribution mechanism, micro-irrigation practices, region-based targeted irrigation technology and end-to-end solution to agricultural practices.

At this juncture, policy frameworks that incentivize the use of precision agriculture, promote the development of smart infrastructure and provide financial support for technology adoption are essential for water management practices. Government collaboration with the private sector can further accelerate this transition. Moreover, initiatives to ensure that farmers and stakeholders have the necessary skills to leverage advanced technologies are crucial. Government agencies, academic institutions and private enterprises should collaborate to explore new technologies, improve existing ones and address challenges unique to the Indian agricultural landscape.

Conclusion

Achieving water security in the agricultural sector is a multifaceted challenge that demands a holistic approach. Advanced technologies present a transformative opportunity to address water scarcity issues and enhance agricultural productivity by embracing the prospects of AI and ML, IoT, digitalization and circular economy principles, ensuring a more sustainable and water-secure agricultural landscape. As we navigate towards the Sustainable Development Goals (SDGs) for 2030, the integration of these technologies into mainstream agricultural practices becomes not just an option but also a necessity for a water-secure and food-sufficient India.

Disclaimer: *Views, thoughts and opinions expressed in the article belong solely to the author.*

Water Security for Agriculture

Meenesh Shah[1]

Introduction

Water is at the core of sustainable development and is critical for socio-economic development, healthy ecosystems and human survival itself. However, increasing population, rapid urbanization and industrialization are exerting pressure on freshwater resources, leading to their depletion and degradation. Water security refers to the availability of adequate and clean water resources to meet the growing demand while ensuring the sustainability of water ecosystems. Water security is a major challenge in India due to over-extraction of groundwater, unequal distribution of water resources, imbalance between demand and supply, lack of proper water management, water pollution, climate change and conflicts over water.

Challenges Faced by the Agriculture Sector

The agriculture sector accounts for the largest share (about 90%) of the total annual freshwater withdrawal in India, which is much higher than its global counterpart (72%). The demand for water in agriculture is driven by a growing need for food due to a rapidly increasing population, the efficiency of crop and water management practices and climatic factors. The heavy reliance on water as an agricultural input in the country is also explained by the preference for cultivating water-intensive crops such as paddy, wheat and sugarcane. About two-thirds of the agriculture irrigation requirement in the country is met through groundwater, leading to groundwater depletion at an alarming rate (0.3 meters per year as per FICCI, 2021). Declining water tables are resulting in increased costs of pumping and irrigation. Declining water quality, such as salty irrigation water as a result of over-extraction, is leading to crop and revenue losses for farmers and could have long-term consequences for water availability. The imbalance between demand and supply of freshwater, coupled with unequal distribution of water resources across the country (some regions

[1]Chairman, National Dairy Development Board (NDDB), Anand, Gujarat.

facing severe water scarcity while others have surplus water), is contributing to increased water stress.

Although our country has one of the world's largest irrigation systems, it is characterized by a high level of inefficient water use. The lack of proper water management systems, leaky water supply networks and insufficient wastewater treatment facilities are further increasing the severity of the problem. Additionally, certain policy loopholes, such as the lack of incentives for efficient and sustainable water use, inadequate water charging, concessional pricing of water pumps, energy subsidies for groundwater pumping and inadequate enforcement and monitoring of existing water policies, would further contribute to the depletion of water resources. Global warming is expected to increase the rate of evapotranspiration and decrease the availability of water for irrigation. Climate change is resulting in erratic rainfall patterns (less frequent but extreme rainy days) and an increased frequency of natural disasters such as floods followed by intense droughts, thereby intensifying land degradation in rainfed agriculture in India. Climate change is significantly increasing the strain on water resources, affecting water availability and agricultural productivity, as about 55% of arable land in our country is heavily dependent on the monsoon (Lee, 2019). Factors such as the imbalance between water demand and supply, poor water resource management and climate change would together contribute to transforming our country from a water-stressed to a water-scarcity position by 2050.

Initiatives for Improving Water Security

To promote sustainable water management and improve water security in the country, the Government of India is implementing various schemes and initiatives. These include the *Jal Shakti Abhiyaan* (groundwater availability), *Pradhan Mantri—Kisan Urja Suraksha evam Utthaan Mahabhiyan* or PM-KUSUM (promoting grid connected solar pumps for irrigation), *Jal Jeevan Mission* (drinking water availability), National Water Policy (rainwater harvesting and conservation), Groundwater Legislation (includes rainwater harvesting), regulation and control of groundwater management through Central Groundwater Authority, Master Plan for Artificial Recharge to Groundwater, *Atal Bhujal Yojana* (sustainable management of groundwater), National Water Mission (Sahi Fasal Campaign), National Water Awards for recognizing efforts in water conservation, Mission Water Conservation (natural resource management through Mahatma Gandhi National Rural Employment Guarantee

Scheme or MGNREGA and *Pradhan Mantri Krishi Sinchayee Yojana* or PMKSY), as well as mass awareness programs, among others.

1. *Sustainable Use of Groundwater through Solar Energy*

Sustainable use of groundwater is crucial for future agricultural security. Recognizing this, the National Dairy Development Board (NDDB) has successfully demonstrated a pilot project on how farmers can sustainably use groundwater for irrigation through solar energy. By using solar pumps, farmers can extract groundwater for irrigation and surplus electricity can be transferred to the grid, generating additional revenue for farmers.

This initiative represents the first solar pump owners' cooperative, where farmers have foregone their claims on highly subsidized energy for irrigation and have instead opted for solar pumps (10 HP and 15 HP). They have established their own micro-grid to sell excess energy generated from the solar panels (Figure 1). Mujkuva SPICE (MSPICE) was inaugurated by Hon'ble Prime Minister of India on 30th of September, 2018 (Figure 2).

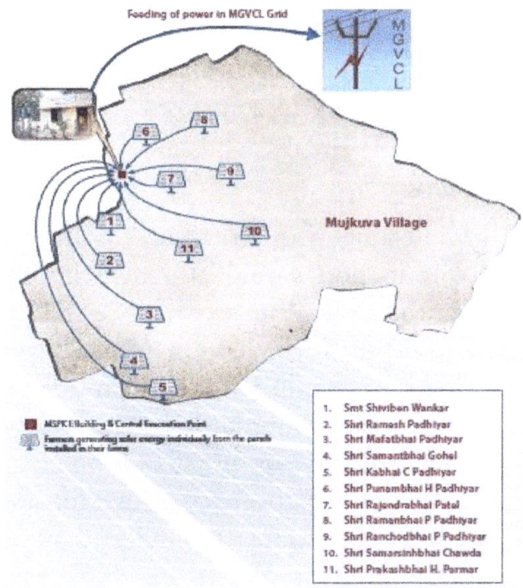

The National Dairy Development Board (NDDB) along with International Water Management Institute (IWMI) and Rajasthan Electronics and Instrumentation Ltd. (REIL) organised 11 farmers in Mujkuva village of Anand district, Gujarat to establish India's first Grid connected Solar Pump Irrigators' Cooperative Enterprise (SPICE) for sustainable use of groundwater for irrigation using solar energy.

Figure 1: MSPICE Network

મુંજકુવા સૌર ઊર્જા સહકારી મંડળીનું ઇ-લોન્ચિંગ કરશે

PM આજે આણંદમાં

અમૂલનાં 6 નવાં પ્રકલ્પોનું લોકાર્પણ, ખાતમુહૂર્ત કરશે

રાજ્ય સરકાર દ્વારા પર્યાવરણને અનુરૂપ ખેડૂતલક્ષી સૂર્યશકિત કિસાન યોજના અમલી

સૂર્ય ઊર્જાથી 11 ખેડૂતોને વીજ બિલથી મુક્તિ મળી, વધારાની વીજ વેચવાથી નફો રળાયો

મુજકુવાના ખેડૂતો સૌર ઊર્જા સહકારી મંડળીના માધ્યમથી સૂર્યશકિતથી સિંચાઇ કરી રહ્યા છે

NDDB અને IWMIના સહયોગથી સોલર પંપ અપાયા

Figure 2: Press Clipping on Launching of MSPICE

This model is helping conserve groundwater, as farmers use the irrigation pumps judiciously to maximize their income from selling solar energy to the grid. Apart from achieving self-reliance (*Atmanirbharta*) in meeting energy needs for irrigation and reducing carbon emissions, the MSPICE farmers are also saving government subsidy on electricity (about Rs. 6.5 lakh per year). The MSPICE model has served as a reference for the *Suryashakti Kisan Yojana* (SKY), a Government of Gujarat scheme for promoting feeder-level grid-connected solar pumps and for designing component C of the PM-KUSUM scheme for grid-connected solar pumps.

2. *Water Use Efficiency of Milk Production*

Apart from efficiently utilizing groundwater for agriculture, optimizing water use efficiency for livestock production would also help improve water security in the country. Within the agriculture sector, about 31% of freshwater withdrawal is attributed to livestock production. Water plays a crucial role in dairying, as the production of milk requires a considerable amount of freshwater (more than 90% as 'indirect water use' for cultivating feeds and fodders and less than 10% as 'direct water use' for drinking and servicing animals). Depending on the source, freshwater use for milk production can be further divided into blue water (ground and surface water) and green water (rainwater) components. Water use efficiency or the water footprint of milk is a measure of the total volume of freshwater consumed in different steps of the milk production chain, expressed as water consumed (litres) per kilogram of milk.

Feeding dairy animals imbalanced rations (without considering critical nutrients such as energy, protein and minerals) often results in lower milk productivity and a higher water footprint of milk. However, if animals are fed balanced rations, comprising a judicious mix of green fodder, dry fodder, concentrate feed ingredients and mineral mixture, it is possible to improve milk productivity and reduce the water footprint of milk. To address this, the National Dairy Development Board (NDDB) conceptualized and implemented the 'Ration Balancing Programme' on a large scale in the country (covering 2.8 million cows and buffaloes in 33,000 villages across 18 states). Feeding animals nutritionally balanced and least-cost rations can help reduce the water footprint of milk by 15% (Figure 3). These findings reveal that, for every one kilogram of fat-and-protein-corrected milk (FPCM) produced, there is a saving of about 155 litres of freshwater (mainly attributed to the indirect water use for feeding). Therefore, there is a substantial scope for reducing the water footprint of milk through the adoption of scientific feeding practices by farmers in the country.

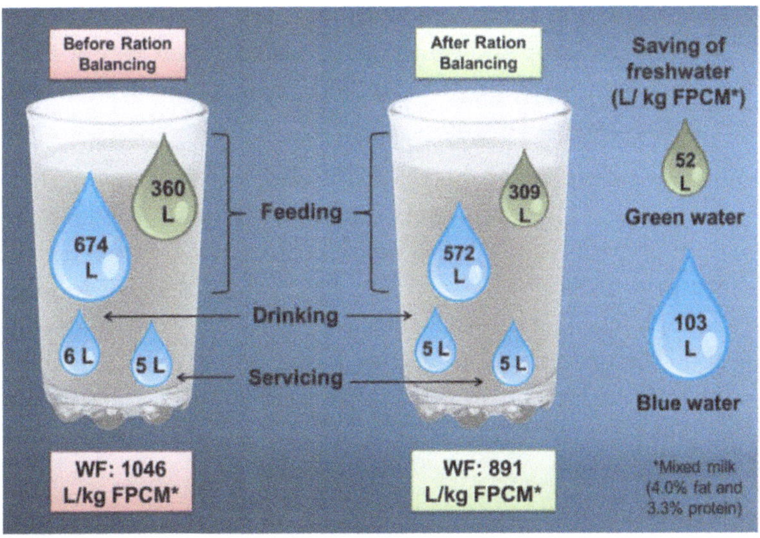

Figure 3: Effect of Feeding Balanced Rations on Water Footprint of Milk

The water footprint of feeds (litres per kilogram of dry matter) varies considerably. In general, crop residues, their by-products and green fodders have a lower water footprint than grains (Figure 4). Therefore, cultivating crops with a lower water footprint and using them judiciously for animal feed would further help reduce the water footprint of milk.

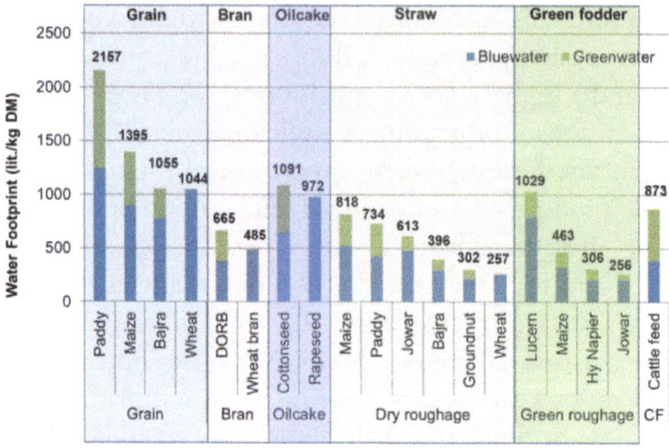

Figure 4: Water Footprint of Feed Ingredients

Additional reductions in the water footprint of milk are possible through the adoption of improved feeding management practices, such as balancing rations using feeds with a low water footprint, delivering season-specific ration advisory services by milk unions or using ration balancing tools such as the e-Gopala android application by farmers, improving the nutritive value of low water footprint crop residues, ensuring year-round availability of fresh and conserved green fodder and feeding Total Mixed Rations according to the stage of lactation.

Agronomy practices can also play a significant role in reducing the water footprint of milk. These practices include adopting good agricultural practices and irrigation techniques such as sprinklers, micro-irrigation and sub-surface drip irrigation, cultivating less water-intensive fodder crops such as Thorn-less Cactus in semi-arid regions, using high biomass-yielding fodder crop varieties or certified seeds, incentivizing green fodder production and conservation and improving rainwater harvesting and watershed management.

3. *Soil Organic Carbon*

It has been scientifically established that increasing the organic carbon content of soil enhances its water-holding capacity (Olness and Archer, 2005). In this regard, the Manure Management initiative by NDDB (Figure 5) is significant. Under this initiative, dairy farmers with a herd size of 3 to 4 animals are supported for the installation of biogas plants in their backyards. These plants provide them with renewable energy (biogas) as well as organic fertilizer (slurry).

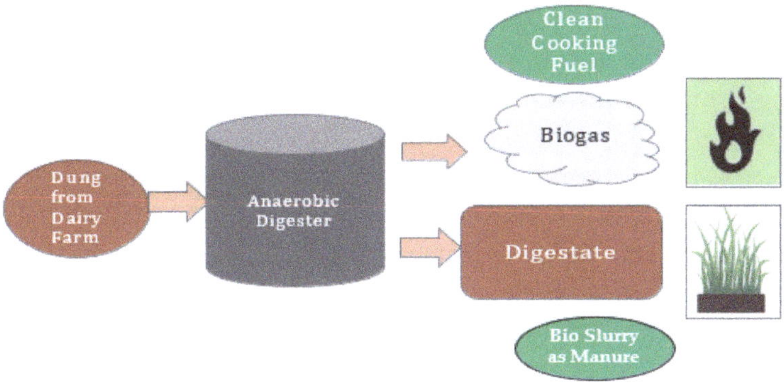

Figure 5: Manure Management Initiative

The slurry is a rich source of organic carbon as well as other micronutrients. Therefore, applying the slurry to farmers' fields results in the improvement of water availability and its quality, besides enhancing soil quality. This initiative is centred around the concept of *Atmanirbhar* farmers, as it reduces their dependency on external stakeholders for access to energy and rich organic fertilizer on a daily basis.

4. *Organic Farming*

Besides the addition of organic carbon, organic farming or regenerative agricultural practices such as cover cropping, crop rotation and minimized tillage have also been shown to improve water availability and quality (Sivaranjani and Rakshit, 2019).

NDDB is the chief promoter of National Cooperative Organics Limited (NCOL, Figure 6). NOCL offers institutional assistance in consolidation, certification, testing, procurement, storage, processing, branding, labelling, packaging and providing logistic facilities for promoting organic farming.

Figure 6: Launch of Logo of NCOL

This support also extends to the marketing of organic products and arranging financial aid for organic farmers through member cooperatives, including Primary Agricultural Cooperative Credit Societies (PACS) and Farmer Producer Organizations (FPOs). Additionally, the organization will undertake all promotional and developmental activities related to organic products in collaboration with various government schemes and agencies. If more farmers can be convinced to practice organic farming through support for marketing their products, it would result in a significant improvement in groundwater availability and quality.

5. *Agroforestry*

It has also been established that increasing tree cover on farms can help recharge groundwater, as the leaf litter and root structure of trees reduce runoff, support water retention and facilitate the trickling of rainwater to groundwater (Ilstedt *et al.*, 2016). Encouraging agroforestry practices among farmers can result in the augmentation of groundwater. To make farmers aware of the importance of trees on their farms and assess their willingness towards agroforestry, NDDB has undertaken an agroforestry pilot. In this pilot, farmers are encouraged to plant various types of fruit trees on their farms and are incentivized through subsidies for the saplings, depending on their survival rate.

Way Forward

Improving India's water security requires a combination of policy, technology and behavioural solutions. One essential strategy is to enhance water governance systems, with clear policies, laws and institutions to manage and allocate water resources effectively and equitably. Efficient water management practices such as rainwater harvesting, conservation and recharge should be adopted to reduce water wastage and increase availability. The country needs to invest in water infrastructure such as dams, reservoirs and canals to improve water storage capacity, distribution and mitigate the impacts of water-related disasters. Promoting the adoption of water-saving technologies like drip irrigation, micro-irrigation and water-efficient appliances would help significantly reduce water consumption and improve productivity. Further, the sustainable use of solar-powered irrigation would support crop and livestock production in the country. Adoption of sustainable feeding practices such as ration balancing and Total Mixed Rations would help improve water use efficiency for milk production. Promoting efficient manure management and regenerative agriculture practices such as improving organic carbon, organic farming and agroforestry would also contribute to improving groundwater availability. Additionally, strengthening the policy framework and enforcing regulations to prevent water pollution, ensuring access to safe freshwater for agriculture and livestock and encouraging behavioural changes for sensible water use would also contribute to improving water security.

Addressing water security requires a collaborative and multi-stakeholder approach, with the active engagement of government, civil society, cooperative and private sectors and farmers. Investing in water security would improve water availability, quality and contribute to the overall socio-economic and sustainable development of the country.

References

[1] FICCI (2021). Creating an ecosystem for increasing water-use efficiency in agriculture. Federation of Indian Chambers of Commerce & Industry. Available at:

[2] https://aspirecircle.org/wp-content/uploads/2022/01/creating-an-ecosystem-for-increasing-water-use-efficiency-in-agriculture-AGRI.pdf

[3] Ilstedt, U., Bargués Tobella, A., Bazié, H.R., Bayala, J., Verbeeten, E., Nyberg, G., Sanou, J., Benegas, L., Murdiyarso, D., Laudon, H., Sheil, D. and Malmer, A. (2016). Intermediate tree cover can maximize groundwater recharge in the seasonally dry tropics. Scientific Reports. 6: 21930. DOI: https://doi.org/10.1038/srep21930

[4] Lee, Y.N. (2019). India may need major rainfalls to reverse its economic slowdown. CNBC. Available at: https://www.cnbc.com/2019/07/10/india-economy-impact-of-monsoon-rain-on-agriculture-sector-farmers.html

[5] Olness, A. and Archer, D. (2005). Effect of Organic Carbon on Available Water in Soil. Soil Science. 172: 90–101.

[6] DOI: https://www.doi.org/10.1097/00010694-200502000-00002

[7] Sivaranjani, S. and Rakshit, A. (2019). Organic Farming in Protecting Water Quality. In: Sarath Chandran, C., Thomas, S. and Unni, M. (eds) Organic Farming. Springer, Cham. DOI: https://doi.org/10.1007/978-3-030-04657-6_1

Integrated Water Resource Management at the Centre of Climate Resilient Agriculture: Some immediate Local Level Considerations & Emerging Trends

Kirit N. Shelat[1] and R. Gopichandran[2]

1. The objective of this narrative is to provide a snapshot of emerging considerations on the water, agriculture and sustainability interface. It highlights India's approach to sustainable development through a systems approach. This effort does not pretend to be scholarly but rather aims to provide a clear and eclectic exercise to help beginners in the field of agriculture studies understand the interrelatedness of Sustainable Development Goals with a special focus on water & agriculture. Sources of important information in this context are presented to facilitate easy access for deeper learning.
2. The centrality of water as part of the energy-food nexus for integrated mitigation and adaptation outcomes to tackle challenges posed by climate change cannot be overemphasized. This is especially true at the water-agriculture interface. SDG 6, focusing on water, interfaces with SDG 2, which addresses food security, with implications for the protection, improvement, use efficiency, supply and restoration of water resources through an integrated ecosystems approach[a-c].
3. India's initiatives are designed to transform the intent of sustainability into reality[d]. To achieve this, India has operationalized a green credit scheme to create a groundswell of locally adapted alternatives for farming aligned with its LIFE initiative[e,f]. Water management and sustainable agriculture are intertwined through this framework, with aspects such as tree plantation and mangrove conservation receiving equal emphasis on resource efficiency. NITI Aayog draws attention to the scope for natural farming[g] supported by an agroecological

[1] IAS (Retd) PhD, DSc, Formerly Principal Secretary, Government of Gujarat. Currently Executive Chairman NCCSD, Ahmedabad (Gujarat).

[2] Professor, NTPC School of Business, Noida (UP).

green villages perspective[h] in the context of SDGs. The water-energy nexus is also equally important in this context[k,l]. NITI Aayog's synthesis of lessons from localizing SDGs is a classic reference[m].

4. Deep insights about adaptation options and barriers across regions with potential impacts on food production are also emphasized recently by the sixth assessment report of the IPCC. It highlights the consequences of water scarcity and insecurity related to observed impacts on major sectors of the economy. Agriculture is the first of these. Solutions could co- evolve with the intensity of locally relevant challenges and therefore, call for a holistic understanding of hydrological cycles, risks and adaptation responses considering qualitative and quantitative profiles of water[n-p]. Its sustained security, store, access, use and preventive measures to reduce vulnerabilities of resources are influenced by seven major enabling[q] principles. These relate to technologies finance, equity and just transitions, culturally dominant use and conservation practices and knowledge, conservation governance and political will to propel and sustain transitions. Deep insights about adaptation options and barriers across regions with potential impacts on food production are also re-emphasized recently by the IPCC[r]. An earlier reference to the dynamics of freshwater resources is a compelling reference in this context. The UNDP presents a bouquet of insights through its compilation on good practices in water security, community-based adaptation, its initiative on Addressing Water Scarcity in Agriculture and the Environment (AWSAMe)[t-w], the contours of financing water adaptive food systems[x] and of women water champions[y].
5. Some of the overarching frameworks in this context include community-based participation for integrated water resources management[z], the specific watershed context[aa], water security[bb], UNDESA's self-sufficient water village aspect based on fourteen actions[cc], along with appropriate rules for partnering[dd], mainstreaming innovations, tackling infrastructure issues, governance, social incentives and monitoring performance[ee]. These frameworks have implications for food and agricultural commodity systems[ff].
6. The present narrative takes note of discussions about the role of water sector-related policies and missions[gg,hh] in India, as referred to in such initiatives as Water for People India[ii], Watershed Organization Trust[jj], the recent excellent snapshot of government and community-led water/watershed programs across the country[kk], initiatives in tribal areas[ll], applications of participatory system dynamics modeling to

help communities visualize water use[mm] and the latest dialogues about climate-resilient water management for sustainable agriculture[nn,oo] from an Indian perspective.

Against the backdrop of the above stated, it is equally important to take additional note of some subtle ground realities with implications for livelihoods for farmers. Three overarching considerations pertain to the fact that water security is critical for a large basket of livelihoods across animal husbandry, manufacturing, infrastructure services, energy production in addition to food security. This calls for a special focus on soil-moisture without which production is not feasible. Issues on this front are complicated further with impacts of climate change at the local level. Farmers are keen to enhance their income; yet constrained by not being able to raise multiple crops due to irrigation related inadequacy in several locations. Farmers totally dependent on irrigation therefore tend to lease agriculture, taking note of diminishing returns. Climate change related rise in sea level affects quality of under-ground water through salinity ingress. While nature's cycle of monsoon is severely affected with recurring droughts. Inclement weather causes unprecedented rains in spells not suitable for periods of planned agriculture. Water along with aspects of soil health are lost through runoffs. This demands well laid out efforts for water security across all periods.

7. 17 locally relevant water–centred facets of sustainable agriculture are accordingly outlined with reference to the above stated. These 17 are at the best indicative and can be embedded in larger frameworks of action. They are also meant to only complement the profile of existing and emerging water-smart policies to serve the agenda of *Atmanirbharta.*

The landscape of Water Smart Policies with the 17 facets relate to:

1. Understanding the water cycle for agriculture is essential for developing effective water management strategies. Even without interference from climate change, underground water resources are depleting due to high discharge from tube wells. This depletion affects both agriculture in rural areas and the management of increasing daily urban water needs. The higher rate of evaporation caused by temperature increases leads to reduced water availability, impacting soil moisture and, consequently, crop productivity and sustainability. These changes also contribute to drinking water scarcity

for cattle and livestock. The rise in sea levels is already being experienced in villages like Dandi-Navsari and Cambay-Anand in Gujarat, resulting in direct losses of agricultural land. Salinity ingress into sub-soil strata affects the root zone, while inundation from floods adversely affects fertile surface soil. Therefore, it is important to understand the water cycle from an alternative perspective and update our understanding annually based on the changes occurring.

WATER CYCLE

Vulnerability of Area – Gujarat

Major Agricultural Systems	*Vulnerability*			*Typical Response Main Climate Option*
	Main Climate Change Exposure	*Sensitivity*	*Adaptive Capacity*	
Kutch	Rainfall variability, droughts, flood	High: mostly rainfed agriculture, marginal lands, poor soil moisture capacity	Low: high prevalence of poverty, limited options, knowledge, social safety nets and resources	Watershed management and on farm water storage for water conservation; Integrated water resources management in river basins; Investment in social infrastructures
North Gujarat	High temperature, rainfall variability droughts	High: crop and animal sensitivity to high temperature and droughts, high population density on marginal lands	Low: high prevalence of poverty, limited options, knowledge, social safety nets and resources limited capacity for water storage	On-farm water storage; crop insurance; increased productivity through better crop-livestock integration; integrated water resources management

Major Agricultural Systems	*Vulnerability*			*Typical Response Main Climate Option*
	Main Climate Change Exposure	*Sensitivity*	*Adaptive Capacity*	
Valsad Navsari South & Middle Gujarat	Reduction in annual rainfall, increased rainfall variability, reduction in runoff and aquifer recharge, high temperatures, higher occurrence of droughts and floods	Variable, depending on the region and level on reliance on agricultural activities. Agricultural systems highly sensitive to changes in temperature and water availability.	Low: adaptive capacity for agriculture in water scarce area	Water conservation where possible; integrated water resources management; crop insurance; improved floods and droughts management plans, shifting out of agriculture.
Junagadh	Increased rainfall variability reduced water availability in places.	Medium to low Some high yielding varieties more sensitive to temperature and water stress, Rain-fed agriculture	Possibilities to compensate water stress through supplemental irrigation in many regions; low capacity in water scarce areas	On-farm storage; for supplemental irrigation; integrated water resources management at river basin level

2. Water harvesting including a focus on check dams, medium and large size dams, ponds, river basin management, de-silting rivers – ponds – dams to enhance storage capacity, recycling of used water, recharging of rain water and used water, Due harvesting and desalination of sea-water;
3. Efficient use of water to learn to live with less, involving drip and sprinkling system, covered canals, transfer by piped system, use water-leak detector, introduce meter for water use – start with nominal charges for all purposes – domestic – agricultural and monitor water transfer losses;
4. ***Water – Energy Nexus:*** With more demand for water in the arena of climate change – more energy will be needed to pump underground

water and transport for new desalination plants and for waste water treatment. Solar energy can be used for desalination of sea-water to enhance water resources and cover open canals with solar panels. It will provide energy and reduce evaporation losses.

5. Coastal areas are affected by overtopping, underground and salinity ingress that affect crop productivity and fish catch. Locally adapted preventive strategies including mangroves-based coast protection, dykes and other nature-based solutions are needed. These could enhance livelihood opportunities and rewards for conservation.

Climate Change and its Impact on Water Cycle

Element of Water Cycle	*Climate Change*
Annual precipitation	Expected to increase globally during the 21st Century, with potentially great spatial variations
Inter annual variation in precipitation	Expected to increase everywhere
Seasonal variability of rainfall	Expected to increase everywhere Delayed monsoon Interim delay within season
Soil moisture stress (droughts)	Moisture stress to generally increase as a result of increasing variability of rainfall distribution (longer periods without rain) and increasing temperatures and deplete soil moisture faster than natural vegetation.
Floods	Increased as a result of increasing frequency and intensity of extreme rainfall events flood intensity can affect standing crops, washing away of upper fertile crust of soil & cause soil erosion
River discharge	Increased variability as a result of changes in rainfall patterns. Changes in annuals runoff expected to vary from region to region.
Groundwater	Varies as a function of changes in rainfall volumes and distribution. Impact is complex with floods contributing to increasing recharge, and droughts leading to increased pumping.
Evapotranspiration	Increases as a function of temperature increases
Water quality (in rivers, lakes and aquifers)	Impact on quality – contamination
Salinity in rivers and aquifers	Potentially high impact where sea water level rise combines with reduced runoff and increased withdrawal
Cyclone	Frequent cyclonic events tornadoes will increase with high intensity.

6. Water Resilient Agriculture: In this background, we need to look at Water Resilient Agriculture and promote same systematically. What is important to know that there are already experiences and expertise available and success stories. But these have limited transfer across geographies and local level adaptation. Some of these experiences are narrated hereinafter.
7. Water Resilient Agriculture in Wetlands: Wetland is an area of land, where soil is saturated permanently or seasonally or covered by shallow water. Wetlands are useful as food source and resource recycling. In terms of products, they are source of fish crops, vegetable and rice crops, medicinal plants and other organic products. The case of East Kolkata Wetlands (EKW) through its bheries/fisheries demonstrates how it caters to the food, sanitation and livelihood security of its nearly a lakh inhabitant by recycling 980 million litres of sewage per day with a detention period of 30 days. About 150 tons of vegetables, 10,500 tons of table fish per day in addition to nearly 11 mt tons of rice per year. This intricate link of eco system productivity based on recycling and livelihood has created a favourable market mechanism that rewards conservation initiatives. These observations are derived from the local leadership of SAFE, with its insights shared on global platforms.

East Kolkata Wetlands

8. ***Water Resilient Agriculture:*** Grasslands, often considered 'common' lands of the community, are crucial yet often neglected ecosystems. They are highly productive but are controlled by none and are

presumed to be the responsibility of no one. Pasture lands, especially in arid and semi-arid regions, are ecologically sensitive and prone to degradation due to the absence of appropriate management practices. Many semi-arid parts of India already have highly degraded pasture lands, leading to high risks of vegetation cover degradation and soil erosion. Recurrent droughts further exacerbate these issues by causing the loss of vegetative cover. During the monsoon season, heavy rainfall on undulating terrain after long dry periods results in significant runoff, washing away large amounts of silt and sand. The thin soil cover in these regions makes them highly susceptible to erosion, rendering the land unproductive after just a few cycles of erosion. To improve the productivity of grasslands, it is essential to implement improved cutting and grazing management practices, remove bushes and other invasive species, rationalize the use of fertilizers and manures and introduce appropriate vegetation, such as legumes, to complement local soil health and enhance productivity.

Appropriate grassland management practices not only contribute to adaptation and mitigation of climate change but also increase productivity, food security and reduce the risk of drought and flooding. Well-managed grasslands provide numerous co-benefits critical to adaptation. Risks associated with prolonged drought periods and unreliable rains can be mitigated by increased water infiltration and retention, which are associated with organic matter accumulation in the soil. This, in turn, improves nutrient cycling, plant productivity and enhances the conservation and sustainable use of habitat and species diversity. Grassland management is thereby a key adaptation and mitigation strategy for addressing Climate Change and variability for livestock and their survival.

9. I*nternational Experience:* CO_2/hectare

10. Water resilient agriculture aligned with agroforestry presents a range of benefits that include protection from soil erosion, facilitates water infiltration, diminishes impact of extreme weather, fodder to livestock – improves soil fertility due to enhanced nitrogen and carbon sinks, hosts for biological control agents and livelihood products. They have to be scaled up significantly. Some images of practices are presented in the following.

A. *Alternate Land Use Systems....... Agro-forestry...!*

B. *Afforestation on Highly Alkaline Soils*

C. *Cultivation of Fruit Trees on Sodic Soil*

Psidium guava

D. *Agroforestry with Saline Water*

Barley with Karonda (*Carissa carandas*)

E. *Control of Water Logging*

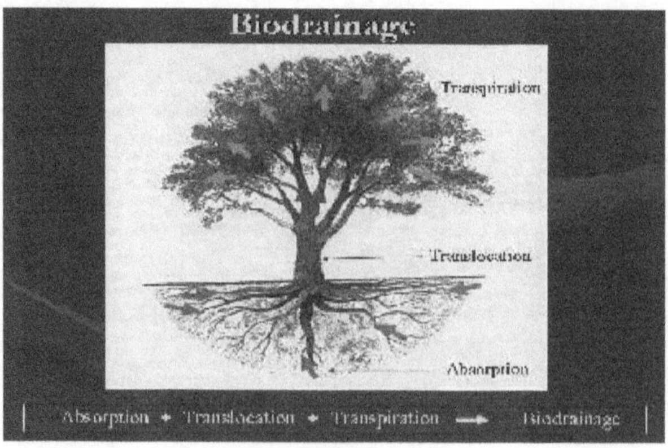

F. *Mangroves: Challenge for Rehabilitation*

11. ***Water Resilient Agriculture:*** Smart Practices – during Adverse Weather Events: Crops have to be protected from the impacts of flow of hot/dry and cold winds through appropriate row and spacing practices Thinning and gap filling must be carried out in the field if rainfall is not received up to 25–30 days after onset of monsoon. Reduce the numbers of rows in the field and adopt inter culturing and weeding operations, apply supplementary irrigation in alternate furrow system in the field, spray the crop with 2% solutions of urea (Hy. Castor, Cotton, pigeonpea) and delay the split application of urea in the crops when rain is delayed. When monsoon is earlier than normal season apply supplementary irrigation to the crop. When monsoon is prolonged than normal, the *rabi* crop like gram can be easily sown on the residual moisture under black soil condition.
12. Some additional aspects that could be adapted pertain to:
    (a) Compartmental block in the field—If slope is observed less than 1% in the field build up the farm bund opposite the slop direction in the field.
    (b) *Contour farming*—If slope is observed 1–2% in the field contour farming must be adopted.
    (c) Strip cropping to prevent the soil erosion in the field—strip cropping cultivation adopted keeping with 6–8 raw in the field and sow the spreading types crop to avoid the soil erosion in the field.

(d) Sow seeds on Ridge and Furrow system deepened up to 15–18 cm and spaced with 30–35 cm apart. Keep 45 distance between two furrows.
(e) Plough in summer season
(f) Form contour/graded bunds in the fields—It is applicable where is slope is observed up to 1 to 6% in the field and total rain fall received more than 800 mm
(g) Promote Silvi-pasture Management and Agri-Horticulture
(h) Panchayats should focus on watershed management and construction of water harvesting structures.

13. Aspects of zero tillage and seaweed cultivation.
14. The case of the Rukmavati river basin management are presented in the following:

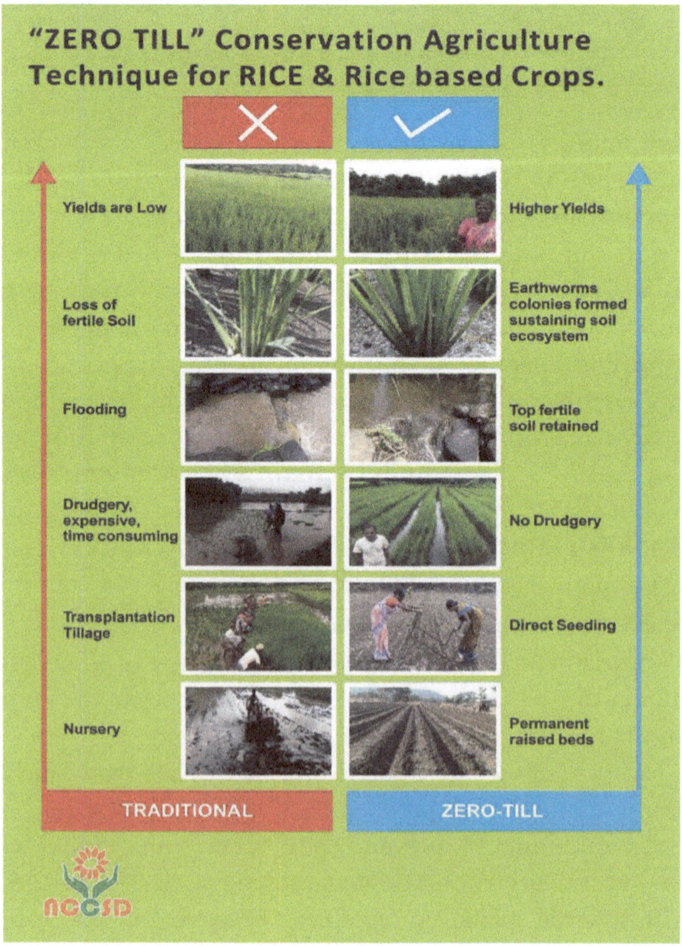

Seaweed cultivation for 'Climate Change Mitigation'

The Concept:

The project aims at establishing & demonstrating following components:

- Seaweed farming unit in coastal area of Kachchh which will serve as raw material source for biomethane generation
- A model power generation unit utilizing biomethane produced though seaweed fed biodigester
- A composting unit producing liquid fertilizer and solid compost from slurry obtained as left over material from biodigester

Project Rationale:

- The seaweed cultivation has potential to provide triple win situation by providing employment opportunities, clean power and higher agricultural production through its use as compost
- It can be developed as a successful cottage or co-operative sector industry. Gujarat has good potential for seaweed farming as it has longest seacoast in India

Seaweed cultivation for 'Climate Change Mitigation'

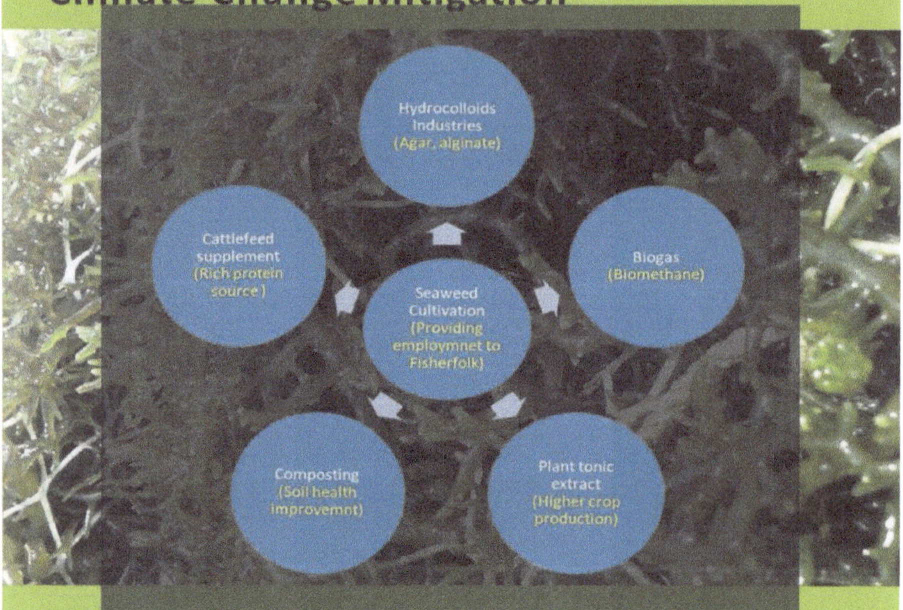

- Increased carbon dioxide in the atmosphere is responsible for rapid climate change
- The ocean provides unlimited space for capturing solar energy by different marine plants through photosynthesis, Seaweed cultivation is one such option
- Seaweed is important not only as part of the marine ecosystem, but also as sources of foods, nutrients and other natural products
- India has a coastal line of around 7,500 km and potential seaweed farming zones can be identified along the coastal states
- Seaweed farming can be considered as a supplementary income source of livelihoods for coastal community including fisherfolks
- Carbon sequestration through seaweed farming. From 1 ton red seaweed (Eucheuma seaweed), the sequestered CO_2 = 1.4 ton/Yr. About 4 ton of wet seaweed can grow in one hectare (ha) area.

NCCSD

Rukmavati River Basin

Casestudy from Kutch, Gujarat

Water Resource Management (Hydrology):

New check dams – 107, Storage capacity - 40 Million Cubic Feet, Beneficial farmers – 1692, Area under irrigation - 3784 hectares

Desilting in 84 structures and 504 hectares benefitting 379 farmers resulted in reduced requirement of chemical fertilizer.

50% saving in water through promotion of drip irrigation in 3684 Ha area.

Natural Resource Management

2000 MT waste converted to compost from 400 Ha of area by 750 farmers resulted in improvement in soil fertility

Promoting legume as inter crop – nitrogen fixation in soil and also providing food security (Nitrogen fixation - 103 MT in 3600 Ha (224 MT urea saving)

Animal husbandry

Pasture land development in 500 Ha of land to provide fodder security.

Farm bunding to conserve soil moisture and control soil erosion, in 64 hectares of land resulted in improvement of productivity by 8 – 10%.

Improving market linkage, 20% increase in income

Tree plantation

35,000 trees planted in basin area to improve greenery

Horticultural plantation of Date palm, Pomegranate, Mango, Ber, Banana and Papaya in 500 Ha to provide income to farmers.

nccsd

Integrated River Basin Management (IRBM)

Burning Issue for the Earth: Climate change and major culprit of it is increasing CO_2 in the atmosphere.

Solution for Mitigating Climate Change: Promoting PHOTOSYNTHESIS to consume CO_2

Landscape Approach for promoting photosynthesis: INTEGRATED RIVER BASIN MANAGEMENT

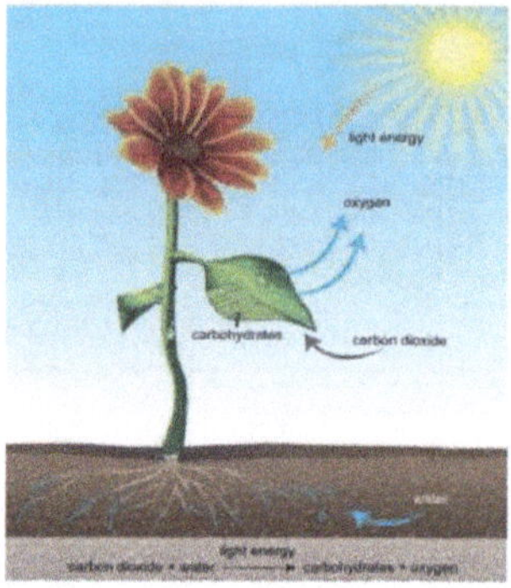

Farms, forests, water bodies and settlement are not isolated elements, but part of a wider landscape in which all land uses are integrated. A landscape approach entails viewing and managing multiple land uses in an integrated manner, considering both the natural environment and the human systems that depend on it.

IRBM is the process of coordinating conservation, management and development of water, land and related resources within the river basin, in order to maximize economic and social benefit while preserving and where necessary restoring freshwater ecosystems.

Better Soil & Water Management → More Vegetation & Green Cover → More Photosynthesis → CO_2 fixation

Soil Health Management → Promoting Waste Recycling for Composting → Better Animal Waste Management → Less Methane Generation

Mitigation leads to Positive Impact on Climate change

NCCSD

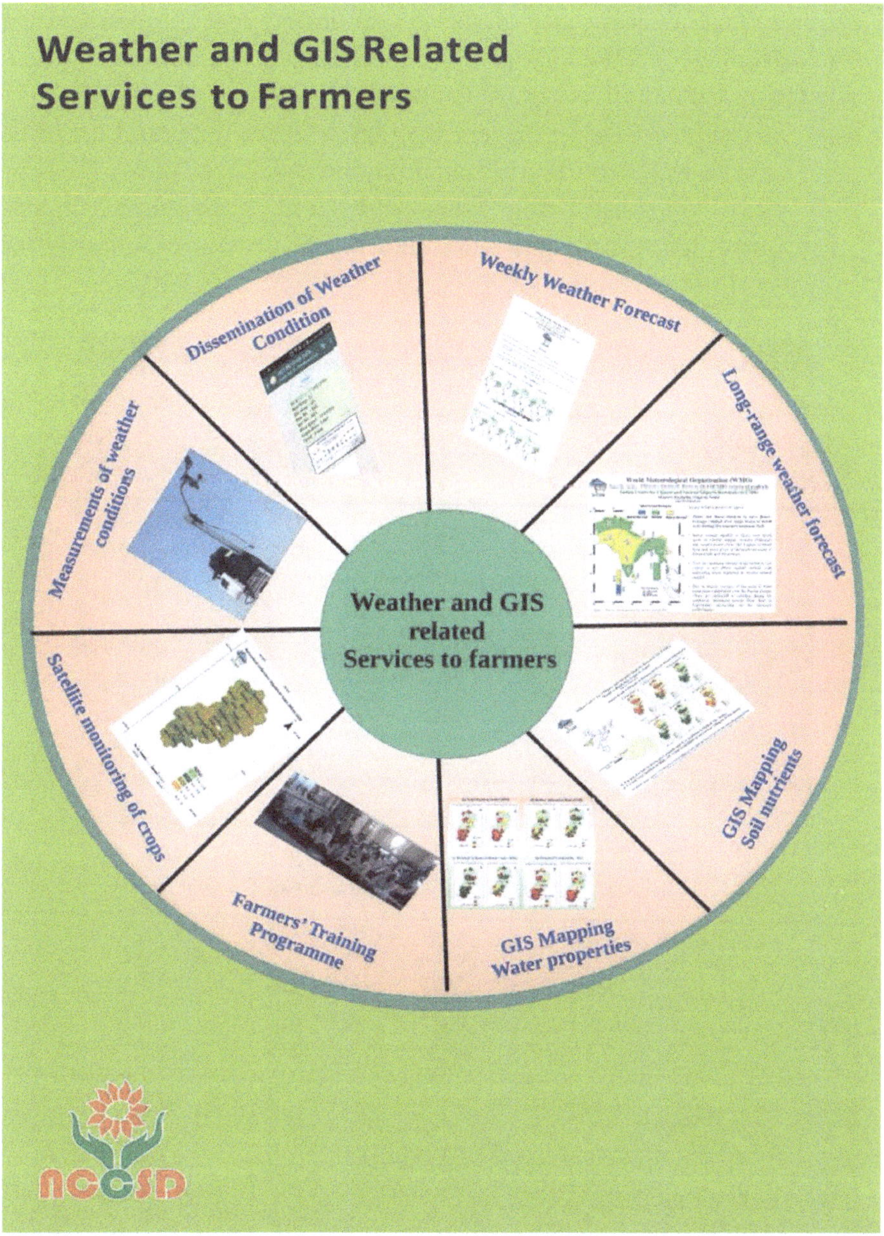

15. ***Water Resilient Agriculture—Management of Saline Water:*** Salinity in Gujarat State is expanding mainly due to i) intrusion of tidal water along the coastal line areas and ingress of sea water, ii) inland salinity due to drought in arid and semi-arid regions, and iii) salinity due to irrigation, irrigation with saline water, poor management of canal,

tributaries and farm water courses in addition to poor agronomic water management practices adopted by the farmers. Related management practices for salt affected soils include:

(a) *Preventive Measures:* Prevent the ingress of sea water and intrusion of tidal water nearby the adjoining low-lying area of sea shore by means of constructing dyke, *pakka* wall, bunding, which can restrict the spreading of tidal water in the interior part nearby the sea shore.

Plantation of mangrove on the sea shore area to prevent the sea water in cultivated land

Prevent intrusion of tidal water in cultivated land nearby sea shore and conserve the rain water by bunding

Planting of trees like bamboos with strip cropping system/pattern around the fields to break the speed of the wind and to protect the field with wind erosion.

Growing of salt tolerant crops like paddy, wheat, sugar beet, mustard, barley and grasses like–Gatton panic, Bermuda grass, Congress grass, Hybrid Napier grass and trees like Mangrove, Nilgiri [Eucalyptus], Saru [Casuarina], Subabul [Leucaena], Vilayati babul [*Prosopis juliflora*], bamboo, etc., which can facilitate to improve the physical, chemical and biological rejuvenation properties of the salt affected soils.

Salt tolerant horticulture crop, Date palm variety Barahee

Ber Plantation in Salt Affected Soils

(b) Control Measures that Incorporate crop residues into the soil to increase soil organic matter (e.g. use of green manure and animal manure). Increased soil organic matter improves soil structure, drainage and aeration.

Green manuring of Sesbania Crop in Salt Affected Soil

(c) Minimum tillage that improves soil structures and in turn improves drainage. Excessive tillage of soils promotes formation of a hard pan below the soil surface (about 0.5 to 1 m deep), causing poor soil drainage.

Broad bed and furrow system of land configuration in salt affected soil

(d) Water Management that is centred on Use of irrigation through Canal – Tube well – In absence of that through water harvesting–water-shed management and even – storing rain water in farm ponds. Irrigate the crop at critical stages of irrigation when there is allow availability of irrigation water. Adopt the micro-irrigation system like drip irrigation system with mulching practices.

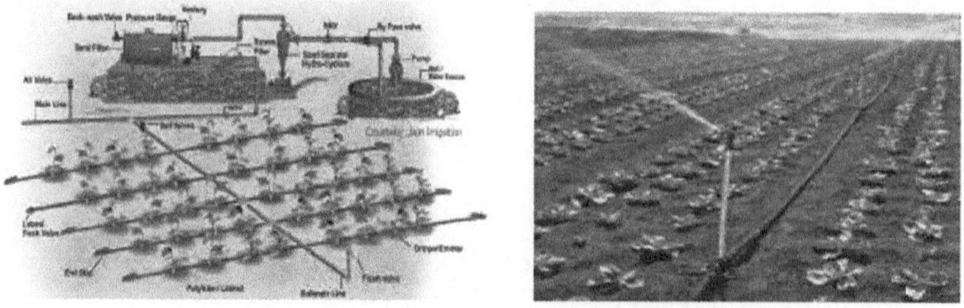

Use of Improved Method of Irrigation
(left: drip irrigation; right: sprinkler irrigation)

(e) ***Soil drainage:*** Make sure that the soil is properly drained to make leaching (washing out) of excess salts possible.

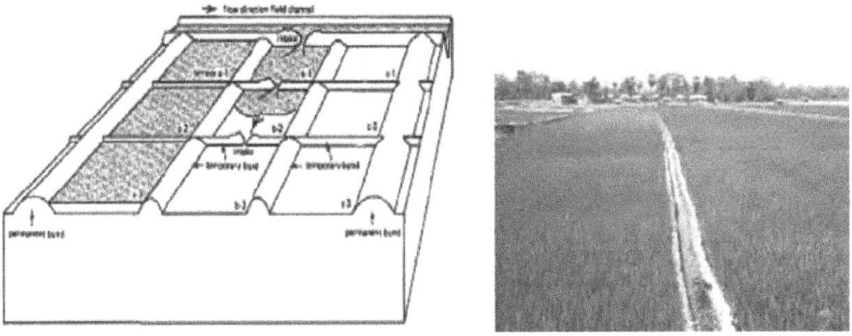

Provision of Drainage System to Reduce Soil Salinity

(f) ***Use of mulch:*** The practices of mulching reduce the evaporation rate and prevent the accumulation of salts from the deep layer of soil on the surface. It can moderate the soil temperature and maintain the soil moisture at optimum level. It can also reduce the infestation of weeds, pests and soil borne diseases.

Use of organic mulch (e.g. crop residues) to reduce evaporation

(g) When soil dry-up in post monsoon season salt tends to reappear on the surface due to upward capillary movement of water. Central Soil Salinity Research Institute, Karnal Studied the effect of rice husk as mulch with or without application of gypsum. It was found that paddy husk mulching checked evaporation from soil surface and prevented upward movement of salts. The yield of paddy increased with increasing depth of rice husk, applied as mulch in plots where no gypsum has been applied.

(h) Use of inorganic mulch i.e. use of plastic mulching. Application of soil amendments:

- Proper and frequent soil testing is required to determine the type and amount of soil amendment needed.
- Use recent/current soil test data to determine the type and amount of amendment to be used.
- Use of fertilizer and fertilizer management should be part of oil management strategy.

Restoration of saline-alkali and alkali soils for growing crops demands that physical properties of soils are improved and excess salts are removed out of root zone. Also, Sodium present in complex of clay must be replaced by clay calcium, which is essential for good structured soils and is an essential plant nutrient. Calcium is generally added as gypsum. Gypsum – $CaSO_4$ is comparatively cheaper than other sources of calcium. Apply gypsum on the basis of soil analysis report. Adopt the heavy green manuring practices of sesbania crop and incorporate bio mass of sesbania when optimum moisture level (nearly 40–45% moisture) available in the soil. Apply the single super phosphate @ 50 kg per hectare for proper decaying of bio mass of sesbania crop.

(i) ***Use of micro-nutrients:*** Experiments at the Central Soil Salinity Research Institute, Karnal have shown that in addition to N, P and K, it is extremely essential to add zinc to these soils. Increased yields up to 20 q/ha of paddy have been obtained by application of $ZnSO_4$—zinc sulphate at the rate of 45 kg/ ha. Zinc sulphate is applied along with other fertilizers by broadcast before transplanting. In subsequent years the dose may be reduced to 25 kg/ha.

(j) Alternative crops—Plant alternative crops that are tolerant (less susceptible) to soil salinity.

(k) Crop rotation with Continuous cultivation of one crop could exacerbate soil salinity; a planned crop rotation could be an alternative of controlling soil salinity & crops with different rooting depth and salinity tolerance.

(l) Plant breeding and bio-technology Selection of suitable crops, crop rotations, method of planting and plant population are very important considerations in reclamation of saline soils and cropping patterns thereafter. During the initial stages of reclamation only salt tolerant crops are taken and as the soils improve sensitive crops can be grown. Adopt the following crop sequences in salt affected soils:

Salt Tolerant Crop	*Medium Salt Tolerant Crop*	*Sensitive Crop*
Field crop: Paddy, Barley, cotton, wheat, mustard, Sesbania, sugar beet, sugarcane	*Field crop:* Oat, paddy, sorghum, millets, maize, gram, cotton, Lucerne	*Field crop:* Field been, cow pea, etc.
Horticulture crop: Date palm, coconut, mangrove, saru, eucalyptus, sisu, acacia, etc.	*Horticulture crop:* Pomegranate, grapes, mango, guava, banana, pich, apple, grape fruit etc.	*Horticulture crop:* Nil
Forage crops: Gatton panic, burmunda, hybrid Napier, sudia jowar	*Forage crops:* Berseem, Lucerne, gatton panik	*Forage crops:* Nil
Vegetable crops: Turpin, spinach, beet root, carrot, garlic, onion etc.	*Vegetable crops:* Tomato, bottle ground, garlic, carrot, beet root, etc.	*Vegetable crops:* Nil

16. Water Resilient Agriculture/irrigation favours multiple approaches and options for all Irrigation Methodologies available for localities in

India combating salinity in source water and soils. The approaches range in level of complexity and costs. Knowledge about the design parameters is paramount. There should be a willingness to experiment with methods, especially with the goal towards finding ways to reduce costs. Irrigation efficiency is described by several terms used to measure how efficiently irrigation water is applied to the field and/or used by the crop. High irrigation efficiency translates into lower operating costs, improved production per unit of water delivered and improved environmental benefit and management and zero transmit loss. Incorrect use of efficiency terms leads to poor performance at village level and reduced number of beneficiaries. Therefore, it is important irrigation department efficiently manage system. Irrigation canals currently face problems of leakages, theft of water by directly lifting from canal and increasing evaporation. One estimate indicates loss of 50% to 60% of water discharged from Dams is lost or wasted. There is no accounting system to measure these losses. In order that maximum farmers' benefit—it is suggested that

(a) Distribution Management separated from Dam Management and account for water made available used.
(b) Prevent slippages and use theft sensors.
(c) Prevent evaporation – covering canals with solar panels for energy on a commercial basis.
(d) Introduce pipe systems and avoid open canals.
(e) Use "Drip System" and flow irrigation system and discourage subsidies.

17. ***Water Resilient Agriculture:*** Development of Wastelands. Due to change in rainfall pattern in last two decades in Gujarat and Rajasthan – drought prone areas are receiving good rains. This is true of desert areas too; wherein the margin areas of desert have become cultivable wasteland. New emerging cultivable wasteland can be used to support a large livestock economy and alternative forestry systems on the basis of scientific evidences from satellite observations. They can then serve the food/fodder/fibre and fuel needs of communities and livestock in the region. These are to build on nature-based solutions, the latest paradigm of judicious use of resources. It is essential to respect nature, take note of the signals she presents and design appropriate development strategies that can be sustained for development and quality of life.

Conclusions

In conclusion, we draw the attention of the reader to some additional most recent insights about the aspects discussed. This includes:

1. Deliberations set for end – February 2024 about drought and integrated basin management planning in large scale systems, water cooperation agreements basins, infrastructure options (https://unece.org/sites/default/files/2024-02/Information%20note_EN_final_clean.pdf); outputs of the "Seventh meeting of the Global network of basins working on climate change adaptation" last year (https://unece.org/sites/default/files/2023-05/Global_network_overview_pilot%20projects_progress_May2023_ENGL.pdf) and another reference to 18 basins as part of a network (https://www.droughtmanagement.info/portal/wp-content/uploads/2023/05/Global-network-of-basins_Water-Convention.pdf) several options for water management for climate smart agriculture at the interface of water sheds/aquifers and river basins (https://www.fao.org/climate-smart-agriculture-sourcebook/production-resources/module-b6-water/chapter-b6-4/ar/), financing for basin – cantered adaptation https://openknowledge.worldbank.org/entities/publication/f038b8cd-0746-5c54-8747-b22a06a05b77 & https://www.impriindia.com/insights/prioritising-climate-finances/); community based watershed management (https://www.sciencedirect.com/science/article/abs/pii/S0959652623041896), lessons aligned with the UN Water theme "Groundwater: making the Invisible Visible" 2022, (https://link.springer.com/collections/cicjgcefaa), the launch of the ICAR's project on climate adaptation atlas for South Asia (https://icar.org.in/project-climate-adaptation-atlas-south-asia-launched-0) NICRA (http://www.nicra-icar.in/nicrarevised/index.php/home1) and deliberations at the most recent conference of the Association of Agrometeorologists and others in India (https://www.agrimetassociation.org/file/17170091383rd-announcement_INAGMET_2024_Final_broucher.pdf);
2. Our recent understanding of wetlands is enriched through a special emphasis on ecological, economic considerations and nature-based solutions; supported by a legal perspective. https://www.climatehubs.usda.gov/hubs/northern-forests/topic/climate-adaptation-tools-wetland-conservation-and-management (https://www.wetlandbasedsolutions.org/nature-based-solutions/; https://iucn.org/resources/conservation-tool/climate-change-resilience-and-adaptation-planning-tool; https://south-asia.wetlands.org/wp-content/uploads/sites/8/2024/02/Sarovar-Vol-9.pdf; https://www.weforum.org/agenda/2023/12/wetlands-carbon-sink-climate-change-mitigation/; https://www.eea.europa.eu/en/newsroom/news/nature-based-solutions-play-crucial-role; https://www.iwmi.cgiar.org/blogs/wetlands-natures-solution-to-climate-change/; https://ehs.unu.

edu/news/news/economics-of-climate-adaptation-eca-the-tool-to-support-countries-and-communities-to-develop-more-ambitious-climate-adaptation-and-mitigation-plans.html; https://climateadaptationplatform.com/climate-adaptation-and-mitigation-through-coastal-wetland-preservation/ https://www.ramsar.org/sites/default/files/documents/library/cop14_18_20_rev1_dr_nbs_eba_e.pdf; https://www.eli.org/vibrant-environment-blog/protecting-wetlands-promoting-nature-based-solutions & https://www.nortonrosefulbright.com/en/knowledge/publications/6ffb0b75/nature-based-solutions) and;

3. The nexus between water – energy and food are also being increasingly recognized and lends itself to integrated mitigation and adaptation practices. Reference to more than forty models sets the context for locally adapted tools that present options across scales of operations integrating information about water, land, finances and energy needs for the systems under consideration requirements, further relating to carbon emissions (https://www.frontiersin.org/journals/water/articles/10.3389/frwa.2022.837316/full; https://www.sciencedirect.com/science/article/abs/pii/S1462901117305403; https://onlinelibrary.wiley.com/doi/10.1002/ird.2800?af=R) with implications for SDGs. (https://unesdoc.unesco.org/ark:/48223/pf0000379588). The need of the hour is to build on the momentum created by the latest dispensation about Atmanirbharta for prosperity. This should be in the form of robust local level action, to present evidences and optimally influence policies for sustainable development. These can also up-scale credible public policy measures and leadership at the local level. This is aligned with the concept Think Global & Act Local.
4. The sources of Important references that guided the narratives are:
   https://www.fao.org/evaluation/list/sdgs/en,
   https://sdgs.un.org/topics/food-security-and-nutrition-and-sustainable-agriculture
   https://www.fao.org/evaluation/list/sdg6/en.
   https://sdgindiaindex.niti.gov.in/#/
   https://pib.gov.in/PressReleaseIframePage.aspx?PRID=1989495
   https://www.indiawaterportal.org/articles/decoding-green-credit-rules-2023
   https://naturalfarming.niti.gov.in/sustainable-development-goals/
   https://www.niti.gov.in/leveraging-agroecological-approaches-for-clean-green-villages
   https://www.niti.gov.in/overview-sustainable-development-goals
   https://www.niti.gov.in/sites/default/files/2023-03/A-New-Paradigm-for-Indian-Agriculture-from-Agroindustry-to-Agroecology.pdf
   https://iopscience.iop.org/article/10.1088/1748-9326/ac08bf/meta
   https://www.unwater.org/water-facts/water-food-and-energy
   https://www.niti.gov.in/sites/default/files/2020-07/LSDGs_July_8_Web.pdf

https://www.cabidigitallibrary.org/doi/full/10.1079/9781800621602.0005,
https://www.worldbank.org/en/topic/climate-smart-agriculture
https://www.fao.org/water/en/
https://www.ipcc.ch/report/ar6/wg2/downloads/report/IPCC_AR6_WGII_Chapter04.pdf
https://www.ipcc.ch/report/ar6/wg2/downloads/outreach/IPCC_AR6_WGII_FactSheet_FoodAndWater.pdf
https://www.ipcc.ch/site/assets/uploads/2018/02/WGIIAR5-Chap3_FINAL.pdf
https://sdgs.un.org/partnerships/addressing-water-scarcity-agriculture-through-partnerships-and-innovation,
https://www.undp.org/sites/g/files/zskgke326/files/publications/CBA%20Community%20based%20adaptation%20to%20CC.pdf
https://www.undp.org/sites/g/files/zskgke326/files/migration/in/good_practices_in_water_security_ideas_for_praxis.pdf
https://sustainabledevelopment.un.org/content/documents/731undp.pdf,
https://sdg.iisd.org/commentary/guest-articles/financing-climate-adaptive-water-resilient-food-systems/
https://nwm.gov.in/sites/default/files/Women%20Water%20Compendium_July%2021.pdf
https://www.preventionweb.net/news/india-community-participation-must-water-management
https://www.sciencedirect.com/science/article/abs/pii/S0959652623041896,
https://unesdoc.unesco.org/ark:/48223/pf0000367904,
https://sdgs.un.org/partnerships/establishment-self-sufficient-water-village-utilizing-all-possible-and-available,
https://sdgs.un.org/partnerships/democratizing-redesigning-rules-partnering-new-models-water-action-sustainable
https://www.frontiersin.org/articles/10.3389/fenvs.2016.00064/full,
https://www.undp.org/facs?gad_source=1&gclid=CjwKCAiAlcyuBhBnEiwAOGZ2S77AGh2PWRLrryQo9n8w1dSMnJR3qJt8CIL6ldZUesrubwFQ7XTSkBoCc2AQAvD_BwE
https://www.teriin.org/sites/default/files/2018-11/water-resource-policies.pdf
https://www.ceew.in/sites/default/files/managing-indias-water-resources-in-sustainable-equitable-ways.pdf
https://waterforpeopleindia.org/community-based-water-resource-management/,
https://wotr.org
https://ies.gov.in/pdfs/water-conservation-coloured-english.pdf

https://idronline.org/article/water-sanitation/how-community-ownership-sustains-water-conservation/
https://www.sciencedirect.com/science/article/abs/pii/S2352801X23000772
https://journalijecc.com/index.php/IJECC/article/view/1894
https://onlinelibrary.wiley.com/doi/abs/10.1002/ird.2696
Additional knowledge resources on wetland conservation and river basin management are:
https://nmcg.nic.in/writereaddata/fileupload/56_Muzaffarnagar.pdf
https://www.sciencedirect.com/science/article/pii/S2351989421001542
https://www.undp.org/india/blog/restoring-and-conserving-wetlands-lesson-community-farming
https://www.sciencedirect.com/science/article/abs/pii/S1364815220310094
https://www.frontiersin.org/journals/water/articles/10.3389/frwa.2023.1059170/full
https://agupubs.onlinelibrary.wiley.com/doi/10.1029/2003WR002936
https://www.undp.org/sites/g/files/zskgke326/files/publications/gef_ridgetoreef_cra_lores.pdf
https://unece.org/sites/default/files/2022-06/Global_network_overview_pilot%20projects_progress_Apr2022_final_ENGL_update%2028.6.2022.pdf
https://wedocs.unep.org/bitstream/handle/20.500.11822/20479/Freshwater_strategy_2017-2021.pdf
https://unepdhi.org/wp-content/uploads/sites/2/2020/06/The-role-of-Decision-Support-Systems-and-Models-in-Integrated-River-Basin-Management.pdf

Water-Resilient Agriculture: A Holistic Approach through Natural Farming

K. Timbadia[1]

ABSTRACT: This study explores Natural Farming practices for enhancing water resilience and sustainability in agriculture. The research emphasizes the pivotal role of earthworms in improving soil moisture retention, infiltration rate and overall water-holding capacity. Additionally, the application of Jeevamrit in natural farming demonstrates significant increases in soil organic carbon, microbial diversity and water-holding capacity. Combining zero tillage with cover crops proves effective in enhancing water availability and improving soil structure, while crop diversification emerges as a key strategy for water conservation. The study highlights the positive correlation between soil organic matter and moisture retention, underscoring the importance of these factors in mitigating drought risks. By prioritizing these natural and regenerative approaches, agriculture can not only adapt to water-related challenges but also contribute to the broader goals of sustainability and ecological balance.

Keywords: *Natural Farming, Organic Carbon, Soil Water Retention, Water Resilience.*

Introduction

Water scarcity is a pressing global concern, particularly in countries like India, which sustains a significant portion of the world's population (17.74%) with a comparatively small share of freshwater resources (4.5%). The annual utilizable groundwater resources in India stand at 436 BCM, with annual extractable groundwater resources totalling 397 BCM, as reported by the Dynamic Ground Water Resources Assessment Report 2022. The report reveals that the annual groundwater extraction for all purposes amounts to 239 BCM. Within this total, a substantial 87%, equivalent to 209 BCM, is allocated for irrigation use, while the remaining 13%, representing 30 BCM, is earmarked for domestic uses. In terms of regional distribution, Uttar Pradesh emerges

[1] Vice Chancellor, Gujarat Natural Farming Science University, Halol, Gujarat-389350, India.

with the highest net annual groundwater availability, with a significant 72 BCM. In contrast, Delhi faces the most challenging scenario, with the least net annual groundwater availability recorded at a mere 0.32 BCM.

Bhullar and Sidhu (2006) outlined the critical state of agriculture in Punjab, attributing it to the excessive depletion of natural resources such as land and groundwater. They emphasized that government policies, particularly the provision of input subsidies for nitrogen fertilizers and electricity, along with the minimum support price (MSP) for crops like paddy and wheat, have incentivized the cultivation of water-intensive crops. This shift in cultivation practices has imposed significant pressure on both soil fertility and groundwater reserves. Additionally, according to the National Commission on Integrated Water Resources Development, the irrigation sector alone is anticipated to require an additional 200 BCM by 2050 compared to the demands forecasted for 2025. The Union Ministry of Water Resources has projected that India's annual water requirement, which stood at 1100 BCM in 2017, is expected to increase to 1447 BCM by the year 2050 (State of India's Environment report- 2023).

The current scenario necessitates a comprehensive and sustainable approach to agricultural practices to ensure water resilience. To address this challenge, water-resilient agriculture has emerged as a crucial concept, emphasizing efficient, sustainable and adaptable irrigation practices. This article explores the role of natural farming as a holistic solution for the looming water deficit.

Natural farming is presented as a chemical-free traditional farming practice that emphasizes agroecology and conservation practices. The pillars of natural farming in irrigated agroecosystems include the use of indigenous cow, minimum or zero tillage, residue management, mulching, vapsa and various natural inputs like *Jeevamrit*, *Ghanjeevamrit* and *Beejamrit*. Adopting Natural Farming practices like no-till, implementing cover-cropping with high-residue crops and diversifying crop rotations are recognized strategies that enhance water holding capacity and promote soil health (Ashworth *et al.*, 2017). This approach promotes biodiversity, resilience to water variations and conservation of depleting natural resources.

Improving Soil Health and Water Retention through Natural Farming Practices

Under the soil, earthworms quietly contribute to water-resilient natural farming. They play an important role in making farming competent to withstand water-related challenges. Studies, such as the ground-breaking work by (Hallam and Hodson, 2021), illuminate the transformative impact of introducing endogenic earthworms, Aporrectodea caliginosa. The research underscores a remarkable 17% increase in soil moisture holding capacity, 27% more available water and a near doubling of infiltration rates over a decade. Simultaneously, the high water-holding capacity of earthworm casts and the interstitial pores between them, as revealed by (Blanchart *et al.*, 1999), contribute to enhanced soil water storage. Moreover, extending beyond their mechanical contributions, as reported by Hajam *et al.* (2023), earthworms aerate the soil, carried out bio-tillage, fostering oxygen flow to plant roots and creating channels for water movement. In the complex interactions of soil, earthworms in natural farming not only enhance immediate water retention but also pave the way for resilient and sustainable agriculture.

The application of Jeevamrit in natural farming has been shown to increase soil organic carbon content, microbial diversity and water-holding capacity. The study by Saharan *et al.* (2023) demonstrates a significant 46% total increase in soil organic carbon due to the application of Jeevamrit over three years, starting at 0.61% in 2017 and reaching 0.92% in 2019. Concurrently, the soil application of Jeevamrit resulted in a remarkable enhancement in microbial diversity, reflected in a microbial count 528 times higher than that in soils from chemical-based farming fields. These findings underscore the dual benefits of Jeevamrit, contributing to both carbon sequestration and a thriving microbial community.

Tillage disrupts soil aggregates, diminishing the essential spaces for air and water movement. Aeration induced by tillage intensifies microbial activity, leading to faster consumption of soil organic matter and resulting in unhealthy soils with compaction and diminished water-holding capacity. The loss of the soil's "sponge" diminishes its ability to store water, posing challenges during severe weather events (Bryant, 2015). Moreover, under zero tillage, West and Post (2022) reported an average soil organic carbon (SOC) sequestration rate of 0.57 ± 0.13 Mg C ha^{-1} $year^{-1}$ (1 Mg = 106 g) up to a depth of 30 cm.

When cover crops are employed in conjunction with zero tillage, they play a pivotal role in enhancing water availability through carbon sequestration, improvement of soil structure, biological activity and reduction of evapotranspiration (Bhadha *et al.*, 2017). Additionally, the utilization of cover crops along with zero tillage, combined with crop residue mulch, further contributes to reduced evapotranspiration and the degradation of organic carbon, as reported by Joyce *et al.* (2002). Moreover, cover crops contribute by fostering dynamic living root systems that establish symbiotic associations with fungi (Steenwerth and Belina, 2008). Various mycorrhizal fungi associate with nearly 90% of global plant species, relying on living roots for sustenance since they are unable to produce their own food. This mutualistic relationship involves the exchange of carbohydrates from living roots to fungi, which reciprocate by supplying nutrients and water from the soil to the roots. The synergy extends to the release of polysaccharides and sugars by fungi, nourishing soil bacteria and establishing a balanced ecosystem that enhances soil health.

Presently, the prevalent practice of burning crop residue is employed for stubble load management, leading to persistent issues such as nutrient loss, air pollution and deterioration of soil health. Additionally, this practice results in a decline in soil organic carbon (SOC), as evidenced by a field trial conducted over 19 years in South-Eastern Australia. The trial reported a substantial loss of 1.75 Mg C ha^{-1} $year^{-1}$ in the 0–10 cm soil layer (Poeplau *et al.*, 2015). According to Singh *et al.* (2016), the application of residue mulch treatments resulted in a noteworthy increase in SOC content from 0.45% to 0.55%. On the contrary, retaining crop residues has been demonstrated to enhance SOC content, particularly in the initial two decades, with benefits diminishing over the long-term (Kirkby *et al.*, 2016). In rainfed and arid regions where water and crop residues are scarce, including those for animal consumption, innovative techniques like plastic mulch in conjunction with drip irrigation and the application of *Jeevamrit* have shown promising results. These practices not only address water scarcity and limited crop residues but also offer a sustainable approach to agriculture.

Crop diversification is recognized as a strategy to improve water resilience in agriculture. By promoting the cultivation of less water-intensive crops through crop rotation, mixed cropping, or intercropping, water resources can be conserved. Montek Singh Ahluwalia emphasized the importance of crop diversification in achieving water sustainability and revitalizing the agriculture sector in Punjab in 2020. The state's goal is to transition 1.2 m ha of paddy out of a total of 3 m ha to alternative crops, with the aim of conserving groundwater. The *Mera Pani-Meri Virasat* Scheme in Haryana exemplifies the implementation of crop diversification to ensure water conservation. The scheme aims to save a substantial amount of water through sustainable agricultural practices.

Soil Organic Matter Content and Soil Water Retention

The impact of anthropogenic climate change is contributing to the decline of soil organic matter (SOM) in agricultural soils, as emphasized by Wiesmeier *et al.* (2016). Soil organic matter serves as a crucial factor influencing soil water retention within the root zone. Consequently, there is an increasing demand to explore the connection between SOM content and soil water retention (SWR). Highlighting the significant impact, Gould (2015) noted that increasing SOM from 1 to 2% could augment the additional water volume to 2.84 litres per cubic foot of soil. Additionally, Sullivan *et al.* (2002) reported that a 1% increase in SOM can hold an additional 1.42 litters of plant-available water in each cubic foot of soil. Scott *et al.* (1986) further noted that the soil's ability to hold an extra 16,500 gallons of plant available water per acre-foot. Moreover, according to the USDA-NRCS, conservative estimates indicate that each 1% increase in soil organic matter allows soils to retain up to 20,000 more gallons of water per acre (Bryant, 2015). In summary, raising the SOM levels in degraded and depleted soils could mitigate drought risks amid climate change and prevent the excessive withdrawal of scarce water resources for supplemental irrigation (Table 1 and Table 2).

Table 1: Positive Responses of Plant Available Water Capacity to an Increase in Soil Organic Matter Content

Increase in PAW Observed for Each Unit Increase in SOM Content	*Reference*
1% rise in organic carbon results in a corresponding increase of 1–4 mm in plant available water	(Wösten and Groenendijk 2019)
Increase in organic carbon content significantly enhances water holding capacity by 1.3–2.3%	(Bhadha *et al.*, 2017)
Increase in organic carbon content significantly enhances water holding capacity by 1.5–1.7%	(Libohova *et al.*, 2018)
Increase in organic carbon content significantly enhances water holding capacity	(Ankenbauer and Loheide 2017)
Increase in organic carbon content significantly enhances water holding capacity by 2 to 7.5%	(Olness & Archer, 2005)
Significantly positive results for water holding capacity in sandy soil with low SOM content	(Rawls *et al.*, 2003)

Table 2: Increase in Water Retention in Soil and Saving of Irrigation by Increasing the Soil Organic Matter Content

Enhance in Water Retention	*Reference*
A 1% increase in soil organic matter content results in a 1.5 times augmentation of plant available water capacity equivalent to 10,800 litres of water per acre at a depth of 15 cm.	(Libohova *et al.*, 2018)
Enhanced water retention due to soil organic matter contributes water-stress free periods by an additional 35 days during the dry summer.	(Ankenbauer & Loheide 2017)

Conclusion

As India faces the challenges of water scarcity and pollution, adopting water-resilient agricultural practices, particularly through natural farming, becomes imperative. In nutshell, the interplay of earthworm activity, *Jeevamrit* application, mulching, zero tillage with cover crops and crop diversification emerges as a holistic approach for enhancing water resilience and sustainability in agriculture. Earthworms contribute significantly to soil moisture retention and bio-tillage, while *Jeevamrit* application fosters soil organic carbon and microbial diversity. Zero tillage combined with cover crops improves water availability and soil structure and crop diversification conserves water resources. Understanding the link between soil organic matter and water retention underscores the importance of increasing SOM levels to mitigate drought risks. Together, these natural farming practices offer a promising pathway to a more water-resilient and sustainable agriculture.

References

[1] Ankenbauer, K.J. and Loheide, S.P. (2016). The effects of soil organic matter on soil water retention and plant water use in a meadow of the Sierra Nevada, CA. *Hydrological Processes*, 31(4), 891–901. https://doi.org/10.1002/hyp.11070

[2] Ashworth, A.J., Allen, F.L., Saxton, A.M. and Tyler, D.D. (2017). Impact of crop rotations and soil amendments on long-term no-tilled soybean yield. *Agronomy Journal*, 109: 938–946.

[3] Bhadha, J., Capasso, J., Khatiwada, R., Swanson, S. and Laborde, C. (2017). Raising Soil Organic Matter Content to Improve Water Holding Capacity, *Sl447/Ss661*,10(5). https://doi.org/10.32473/edis-ss661-2017

[4] Blanchart, E., Alegre, J., Duboisset, A., Lavelle, P. and Brussaard, L. (2004). Effects of earthworms on soil structure and physical properties. *Earthworm Management in Tropical Agroecosystems*, 149–171.

[5] Bryant, L. (2015). "Organic Matter Can Improve Your Soil's Water Holding Capacity." NRDC Expert Blog. https://www.nrdc.org/bio/lara-bryant/organic-matter-can-improve-your-soils-water-holding-capacity

[6] Bhullar, A.S. and Sidhu, R.S. (2006). Integrated Land and Water Use: A Case Study of Punjab. *Economic & Political Weekly*, 41(52): 5353–5357.

[7] Dynamic Ground Water Resources of India (2022) Central Ground Water Board, Dept. of Water Resource, River Development and Ganga Rejuvenation, Ministry of Jal Shakti, GoI, Faridabad.

https://cgwb.gov.in/cgwbpnm/public/uploads/documents/168621599595 0835934file.pdf

[8] Gould, M.C. (2015). Compost increases the water holding capacity of droughty soils. Michigan State University Extention, Drought Resources.

https://www.canr.msu.edu/news/compost_increases_the_water_holding_ capacity_of_droughty_soil

[9] Hajam, Y.A., Kumar, R. and Kumar, A. (2023). Environmental waste management strategies and vermi transformation for sustainable development. *Environmental Challenges*, 13: 100747.

[10] Hajam, Y.A., Kumar, R. and Kumar, A. (2023). Environmental waste management strategies and vermi transformation for sustainable development. *Environmental Challenges*, 13(1): 100747.

[11] Hallam, J. and Hodson, M.E. (2021). Earthworms and soil water regulation: A review. *AFRIMED AJ*, 13: 89–135. https://doi.org/ https://doi.org/ 10.1016/S00167061(03)00094-6

[12] Hussain, S., It's time to make Punjab agriculture great again. But how to do so? The Wire, 1 September 2020; https://thewire.in/agriculture/punjab-agriculture-reforms-roadmap-montek-panel.

[13] Joyce, B.A., Wallender, W.W., Mitchell, J.P., Huyck, L.M., Temple, S.R., Brolnstrom, P.N. and Hsaio, T.C. (2002). Infiltration and soil water storage under winter cover cropping in California's Sacramento Valley. *Trans. ASAE* 42(2): 315–326.

[14] Kirkby, C.A., Richardson, A.E., Wade, L.J. Conyers, M. and Kirkegaard, J.A. (2016). Inorganic nutrients increase humification efficiency and C-sequestration in an annually cropped soil. *PLoS One*. 11: e0153698.

[15] Libohova, Z., Seybold, C., Wysocki, D., Wills, S., Schoeneberger, P., Williams, C., Lindbo, D., Stott, D. and Owens, P.R. (2018). Re-evaluating the effects of soil organic matter and other properties on available water-holding capacity using the National Cooperative Soil Survey Characterization Database. *Journal of Soil and Water Conservation*, 73(4), 411–421. https://doi.org/10.2489/jswc.73.4.411.

[16] Olness, A. and Archer, D. (2005). Effect of Organic Carbon on Available Water in Soil. *Soil Science*, 170, 90–101. https://doi.org/10.1097/00010694-200502000-00002

[17] Poeplau, C., Katterer, T., Bolinder, M.A., Borjesson, G., Berti, A. and Lugato, E. (2015). Low stabilization of aboveground crop residue carbon in sandy soils of Swedish long-term experiments. *Geoderma*, 237: 246–255.

[18] Rawls, W.J., Pachepsky, Y.A., Ritchie, J.C., Sobecki, T.M. and Bloodworth, H. (2003). Effect of soil organic carbon on soil water retention. *Geoderma*, 116(1), 61–76.

[19] Saharan, B.S., Tyagi, S., Kumar, R., Vijay, O.H., Mandal, B.S. and Duhan, J.S. (2023). Application of Jeevamrit Improves Soil Properties in Zero Budget Natural Farming Fields. Agriculture, 13, 196. https://doi.org/10.3390/agriculture13010196

[20] Scott, H.D., Wood, L.S. and Miley, W.N. (1986). Long-term effects of tillage on the retention and transport of soil water (Publication 125). Arkansas Water Resources Research Centre, *University of Arkansas, Water Resources Center, Fayetteville, AR. PUB125*. 45.

[21] Singh, V.K., Dwivedi, B.S., Singh, S.K., Majumdar, K., Jat, M.L., Mishra, R.P. and Rani, M. (2016). Soil physical properties, yield trends and economics after five years of conservation agriculture-based rice-maize system in North-Western India. *Soil and Tillage Research*. 155: 133–148.

[22] State of India's Environment Report (2023). *Centre for Science and Environment*. https://www.cseindia.org/state-of-india-s-environment-2023-11601

[23] Steenwerth, K. and Belina, K.M. (2008). Cover crops enhance soil organic matter, carbon dynamics and micro-biological function in a vineyard agroecosystem. *Applied Soil Ecology*, 40: 359–369.

[24] Sullivan, P. (2002). Drought Resistant Soil. In ATTRA (Appropriate Technology Transfer for Rural Areas) National Center for Appropriate Technology. Issue November 2002, p. 7). http://agwaterstewards.org/wpcontent/uploads/2016/08/Drought_Resistant_Soils_ATTRA.pdf

[26] West, T.O. and Post, W.M. (2002). Soil organic carbon sequestration rates by tillage and crop rotation: A global data analysis. *Soil Science Society of America Journal*, 66:1930–1946.

[27] Wiesmeier, M., Poeplau, C., Sierra, C.A., Maier, H., Frühauf, C., Hübner, R., Kühnel, A., Spörlein, P., Geuß, U., Hangen, E., Schilling, B., von Lützow, M. and Kögel-Knabner, I. (2016). Projected loss of soil organic carbon in temperate agricultural soils in the 21st century: Effects of climate change and carbon input trends. *Scientific Reports*, 6(1), 32525.

[28] Wösten, H. and Groenendijk, P. (2019). Soil organic matter and its importance for water management. In Deltafact (Deltafact).

Status and Issues of Water and Food Securities in Gujarat

V.P. Chovatia[1] and H.D. Rank[2]

ABSTRACT: The availability of water resources and its demands from various competing sectors, namely agriculture, horticulture, domestic, livestock and industry, for water security as well as agricultural production and food demands of current populations for food security, were assessed and discussed for the Gujarat state. The state has a cropped area of 99.53 lakh hectares under agriculture and 19.16 lakh hectares under horticulture. The irrigation water demands from the agriculture and horticulture sectors are 42.45 lakh hectare-meters (ha-m) and 7.07 lakh ha-m, respectively. The livestock population and its water requirements are 499.49 lakh ha-m and 0.82 lakh ha-m, respectively. The total domestic water requirements for Gujarat state are 2.779 lakh ha-m for the total human population of 716 lakh. The total water resources of Gujarat include groundwater of 24.58 lakh ha-m and surface water of 9.157 lakh ha-m from the Sardar Sarovar Project (SSP), as well as 19.99 lakh ha-m from various dam reservoirs other than SSP. The share of groundwater and surface water in total water resources is 45.75% and 54.25%, respectively. The total water resources available and water demands in the state are 53.73 lakh ha-m and 54.76 lakh ha-m, respectively, resulting in a deficit of 1.02 lakh ha-m in Gujarat state. The water resources are deficit by 1.88% as compared to total water demands.

Gujarat, despite being a surplus state in food production, faces challenges in water and food security due to reliance on rainfed farming, erratic monsoons and inefficient irrigation. Groundwater withdrawals worsen soil degradation and seawater intrusions. Farmers struggle with volatile market prices, limited technology access and debt burdens. Enhancing water and food security in the state should involve investing in water harvesting, water transfer by river linking, efficient irrigation water management and preventing seawater intrusion. Adoption of climate-resilient crops and organic farming practices is also essential. Strengthening market access, implementing fair pricing mechanisms and establishing social safety nets are crucial for farmer well-being.

[1] Vice Chancellor, Junagadh Agricultural University, Junagadh (Gujarat).

[2] Professor & Head, Department of Irrigation & Drainage Engineering, JAU, Junagadh.

Infrastructural development should focus on fallow/wastelands to retain per capita availabilities of land.

Introduction

The present economy of India is \$3.732 trillion, with contributions of 15% from agriculture, 27% from industry and 58% from the services sector. The Honourable Prime Minister of India has set a target to raise the Indian economy from \$3.732 trillion at present to \$5 trillion by the year 2026–27. Among these, the contribution of the agricultural sector would be raised to \$0.75 trillion from \$0.56 trillion through technological interventions in the agriculture sector, aiming to achieve water and food security. These goals can only be achieved through efficient management of every pixel of land and drop of water resources in each corner of the country, as the per capita availabilities of land and water are decreasing every year.

The annual rainfall in the region varies from 625mm to over 1500mm, with an average annual rainfall of 972 mm. Approximately 24.94% of the area falls under arid conditions and 35.56% falls within semi-arid regions. Gujarat exhibits variations in topography and experiences wide fluctuations in annual rainfall. The rainfall in Gujarat follows an erratic and uneven pattern, resulting in disparities in water distribution among different regions. Since 2005, there have been some changes in these conditions attributed to climate change. Half of Gujarat's districts are categorized as drought-prone. Gujarat's rainfall presents a diverse picture, with significant variations across its regions as depicted in Table 1.

The major *Kharif* crops cultivated in the state include cotton, groundnut, sesame, castor, paddy, pearl millet, maize, pigeonpea, greengram and sugarcane. On the other hand, major *Rabi* crops comprise wheat, rice, maize, mustard, gram, groundnut and pearl millet, among others. Gujarat's land and water resources deviate from the national average. Despite covering 5.98% of the country's total geographical area (328 m ha), Gujarat possesses only 4.8% of the national water resources (1123 billion cubic meters). This comprises 29.15 billion cubic meters (BCM) and 24.58 BCM as surface and groundwater, respectively. The irrigated area in Gujarat is limited to 50.01 lakh hectares, utilizing both surface and groundwater, covering only 44.65% of the total cultivable land.

Table 1: Soil Types and Rainfall in Different Agro-Climatic Zones of Gujarat State

Sr. No.	*Agro-climatic Zone*	*Type of Soil*	*Rainfall (mm)*
1.	South Gujarat (Heavy Rain Area)	Deep black with few patches of coastal alluvial, laterite and medium black	>1500
2.	South Gujarat	Deep black clayey	1000–1500
3.	Middle Gujarat	Deep black, medium black to loamy sand	800–1000
4.	North Gujarat	Sandy loam to sandy	625–875
5.	Bhal & Coastal Area.	Medium black, poorly drained and saline	625–1000
6.	South Saurashtra	Shallow medium black calcareous	625–750
7.	North Saurashtra	Shallow medium black	400–700
8.	North West Saurashtra Zone.	Sandy and saline	250–500

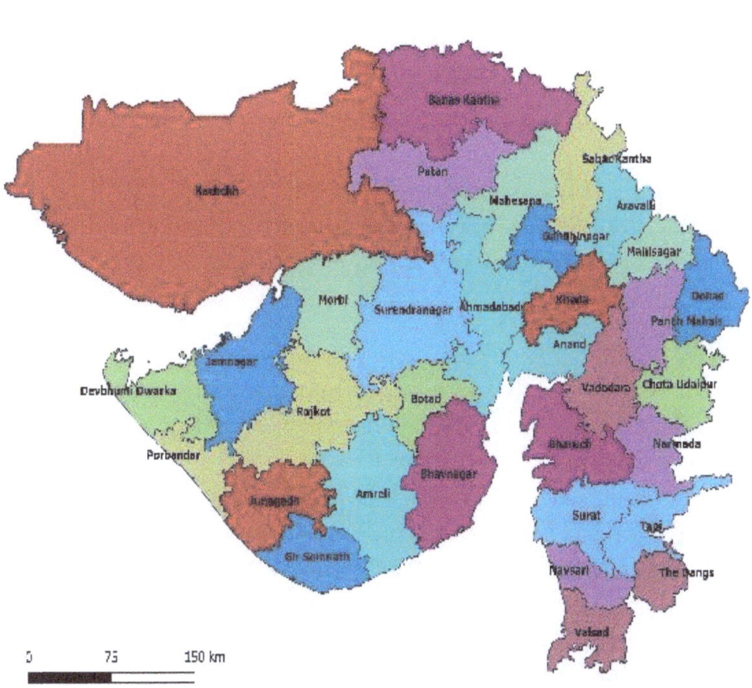

Figure 1: Gujarat State

The overall irrigation potential stands at 64.86 lakh hectares, representing 60% of the state's cultivable land. This includes 39.40 lakh hectares from surface water resources, encompassing major, medium and minor projects and 25.46 lakh hectares from groundwater sources, such as private wells. Gujarat has successfully created irrigation potential for 46.28 lakh hectares, equivalent to 71.33% of the ultimate potential. However, the actual utilization of this potential may vary and could be lower.

Achieving the full potential is contingent on factors such as project completion, effective water management and resource allocation. Challenges such as groundwater overuse and the impact of climate change further complicate this endeavour. The completion of major projects like the Sardar Sarovar Project (SSP) is anticipated to significantly contribute to enhancing irrigation potential, particularly from surface water sources.

Land Resources

The state's total geographical area is 196 lakh hectares; though as per the Survey of India map, it is only 186 lakh hectares. The cultivable land of the state is 123.88 lakh hectares, constituting approximately 63% of the total. Gujarat has a diverse range of soil types, including deep black, sandy, saline, medium black and poorly drained saline alkaline soils. Notably, the state has the largest expanse of saline soil, covering 16.80 lakh hectares and alkaline soil, covering 5.41 lakh hectares. The per capita land availability in Gujarat experiences a continuous decline from 0.827 ha in 1951 to 0.141 ha in 2060. Subsequently, it will rise to 0.154 ha in 2100 (Figure 2). These fluctuations may be attributed to changes in population dynamics and the availability of land for the agricultural sector.

Water Resources

The per capita water requirement for humans, encompassing domestic use and the production of essential items like cereals, pulses, fruits, vegetables, spices and milk, is estimated to be 1,500 m^3. In Gujarat, per capita water availability has exhibited a consistent decline from 3,131 m^3 in 1951 to 770 m^3 in 2024. It is projected to 585 m^3 in 2060 followed by a subsequent increase to 666 m^3 in 2100 (Figure 3). These fluctuations can be attributed to changes in population dynamics and the availability of freshwater for various human purposes. Overexploitation of groundwater and industrial development have significantly reduced the availability of good-quality water. Despite these challenges, the Government of Gujarat has initiated a strategic programme to combat water scarcity. The government is committed to doubling the per capita water availability by 2047. The current per capita water availability of 770 m^3/year in Gujarat is projected to increase to 1,700 m^3 by 2047 (MoWR, GoG, 2023). The analysis of the state's land and water resources underscores the necessity to enhance productivity per unit of land and maximize water use efficiency to meet the growing food demands of the population. Achieving this goal requires technological interventions in the management of land and water resources.

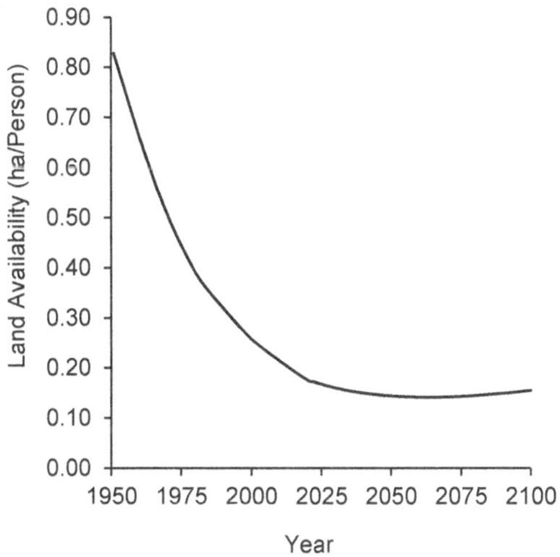

Figure 2: Per Capita Availability of Land in Gujarat

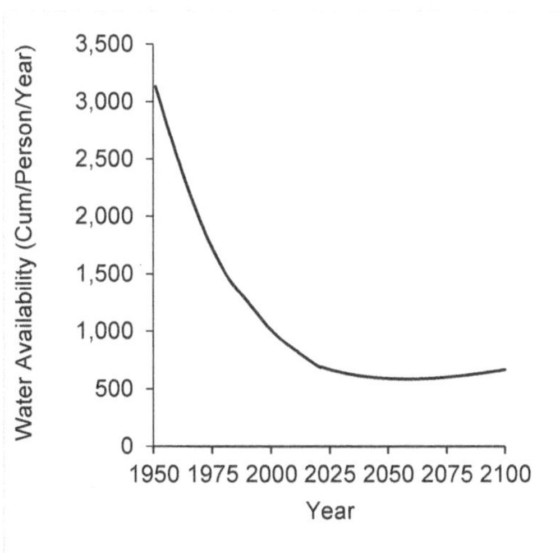

Figure 3: Per Capita Availability of Water in Gujarat

Micro Irrigation Systems (MISs)

The state has the potential to increase its irrigation capabilities by enhancing the coverage under water use efficient irrigation systems like Micro Irrigation Systems (MISs). Presently, MIS covers only 21.83 lakh hectares in Gujarat, indicating a substantial opportunity for expanding irrigated areas through the widespread adoption of MIS (GGRC, 2024).

Until the year 2022–23, a total of 13.87 lakh beneficiary farmers in Gujarat have adopted MISs, covering a cumulative area of 21.83 lakh hectares (Figure 4). Within this total area, 18.21 lakh hectares are dedicated to agricultural crops, while 3.62 lakh hectares are designated for horticulture (Figures 4–7). The widespread adoption of MISs across the 21.83 lakh hectares results in an annual water saving of 3275 MCM, paving the way for expanding irrigation to an additional 6.55 lakh hectares in Gujarat. Furthermore, water harvesting initiatives in the state have fortified surface and groundwater resources, potentially facilitating cultivation in an additional 12 lakh hectares.

The estimated horticultural crop area in Gujarat has experienced significant growth, surging from 5.89 lakh hectares in 1998–99 to over 18.90 lakh hectares in 2023. Given the potential for lower financial inputs, it is advisable to prioritize the coverage of horticultural crops under MIS. However, despite the potential, only 3.62 lakh hectares of horticultural crops currently utilize MIS (Figure 5).

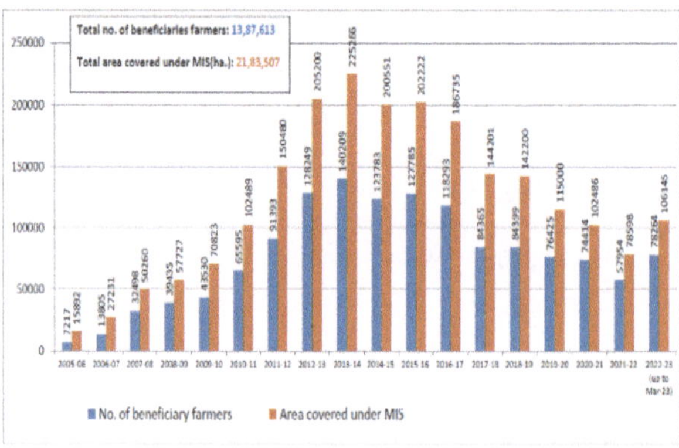

Figure 4: MIS Coverage Acievements (21.84 lakh ha) in Gujarat (2005–06 to 2022–23)

The state's ultimate irrigation potential, combining surface and ground-water resources, is estimated to be 64.86 lakh hectares. Excluding rice area of 9.01 lakh hectares (DoA, 2024), the potential area under MIS is 55.85 lakh hectares. Achieving this potential would increase the total irrigation potential with MISs to 72.605 lakh hectares, considering only water-saving benefits. Additionally, the adoption of MISs could lead to enhanced yield and product quality, energy savings and reduced cultivation costs.

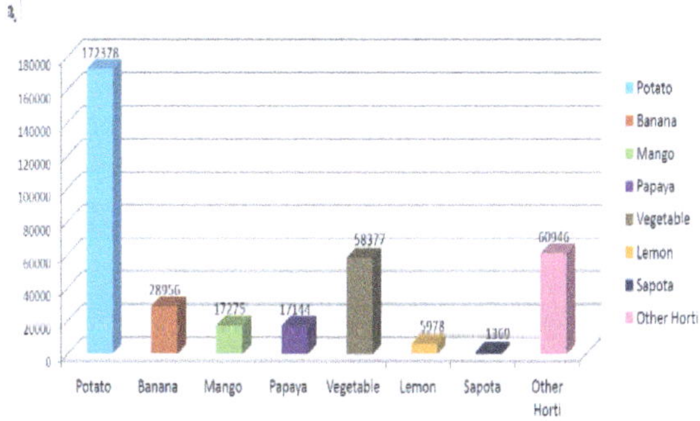

Figures 5: Horticulral Crop Area (3.62 lakh ha) under MIS (2005–06 to 2022–23) in Gujarat

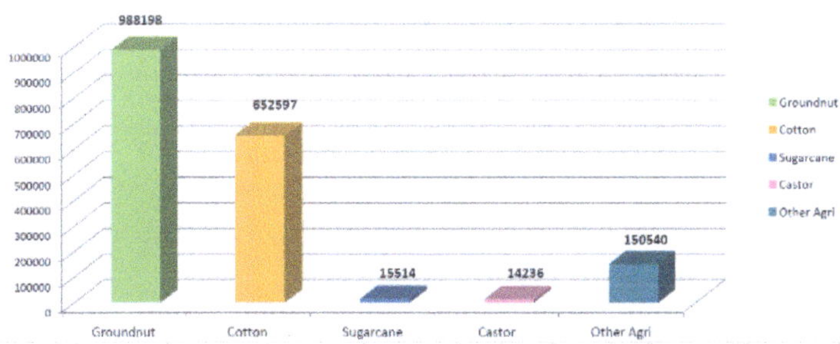

Figure 6: Agricultural Crop Area under MIS Gujarat (2005–2021)

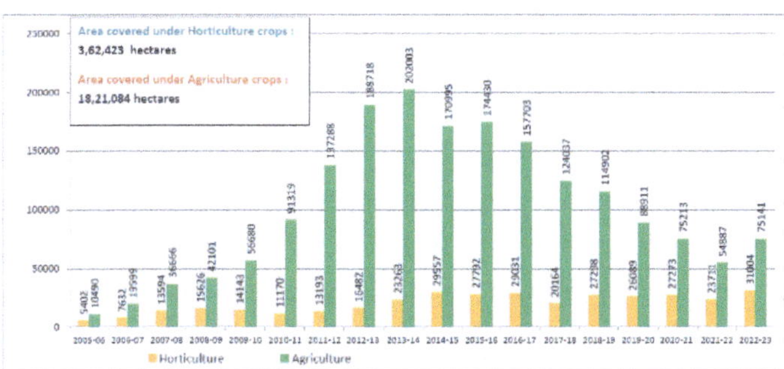

Figure 7: The Agricultural and Horticultural Crops under MIS in Gujarat (2005–2021)

Water Securities

North Gujarat Zone

The north Gujarat zone (Figure 8) has total geographical area of 30.44 lakh ha which includes total 6 districts like Aravalli (3.26 lakh ha), Banaskantha (10.73 lakh ha), Gandhinagar (2.2 lakh ha), Mehsana (4.39 lakh ha), Patan (5.78 lakh ha) and Sabarkantha (4.14 lakh ha). The total water supplies and demands from various sectors in the districts of the North Gujarat zone depicted in Table 2. Among the districts, Banaskantha stands out with the highest total water resources at 2.48 lakh ha-m, inclusive of surface water allocation from the SSP and groundwater resources. Following closely are Mehsana with 1.76 lakh ha-m, Patan with 1.63 lakh ha-m, Aravalli with 1.25 lakh ha-m, Sabarkantha with 1.03 lakh ha-m and Gandhinagar with 0.652 lakh ha-m.

Figure 8: North Gujarat Zone

The North Gujarat region's total water resources comprise 4.25 lakh ha-m from groundwater, 2.24 lakh ha-m from SSP allocation and 2.32 lakh ha-m from surface water resources other than the SSP. Groundwater resources and surface water resources contribute 48.3% and 51.7%, respectively, to the overall water resources in the region.

The highest total water demand from various sectors, including agriculture, horticulture, livestock, domestic and industry is observed in Banaskantha district at 3.56 lakh ha-m. This is followed by Mehsana with 1.72 lakh ha-m, Sabarkantha with 1.56 lakh ha-m, Patan with 1.17 lakh ha-m, Aravalli with 1.06 lakh ha-m and Gandhinagar with 1.00 lakh ha-m.

Among the 6 districts in the North Gujarat Zone (Figure 8), three have surplus water resources, while the other three face a shortfall. Patan district has the highest surplus water resources of 0.4532 lakh ha-m (38.46%), followed by Aravalli with 0.1822 lakh ha-m (17.07%) and Mehsana with 0.043 lakh ha-m (2.50%). Conversely, Banaskantha has the highest shortfall of water resources at –1.08 lakh ha-m (–30.33%), followed by Sabarkantha with –0.533 lakh ha-m (–34.08%) and Gandhinagar with –0.355 lakh ha-m (–35.28%).

In nutshell, the total water resources available in the North Gujarat Zone are 8.81 lakh ha-m, as against the total water demands of 10.1 lakh ha-m, resulting in a shortfall of –1.29 lakh ha-m. This represents a deficit of –12.77% compared to the total water demands.

Table 2: Water Budgeting for the Different Districts of the North Gujarat Zone, Gujarat State

District	*Total Surface and Ground Water Resources (ha-m)*	*Total water Demands (ha-m)*	*Water Surplus (+)/ Deficit (–) (ha-m)*	*Surplus (+)/Deficit (–) of Requirements (%)*
Aravalli	124959	106741	18219	17.07
Banaskantha	248172	356198	–108026	–30.33
Gandhinagar	65192	100733	–35542	–35.28
Mehsana	176435	172130	4304	2.50
Patan	163159	117843	45317	38.46
Sabarkantha	103004	156254	–53250	–34.08
Total North Gujarat	880921	1009899	–128979	–12.77

Central Gujarat Zone

The Central Gujarat zone (Figure 9) covers a total geographical area of 30.09 lakh hectares, encompassing eight districts viz; Ahmedabad (6.97 lakh ha), Anand (2.76 lakh ha), Chhota Udepur (3.46 lakh ha), Dahod (3.66 lakh ha), Kheda (3.39 lakh ha), Mahisagar (2.51 lakh ha),

Panchmahal (3.29 lakh ha) and Vadodara (4.05 lakh ha). Table 3 provides an overview of the total water supplies from various sources and the total demands from different sectors in those districts.

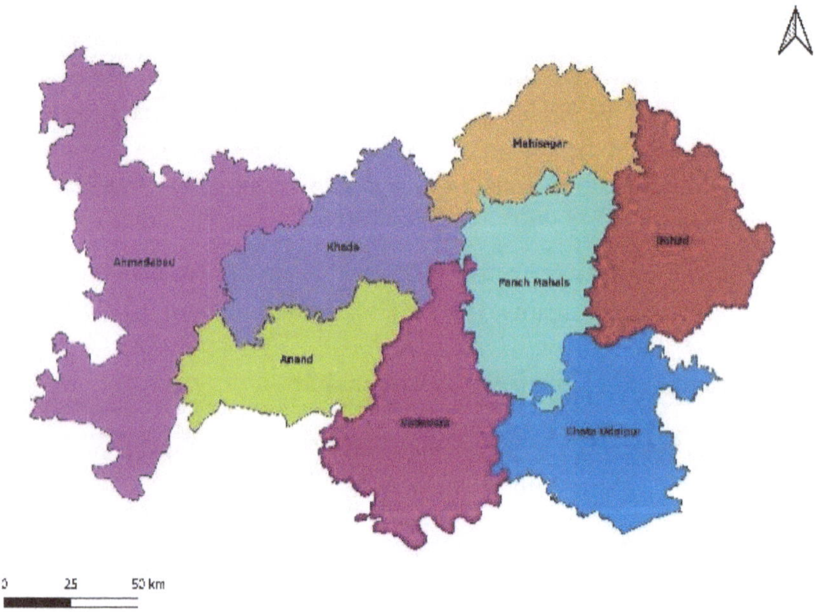

Figure 9: Central Gujarat Zone

The district-wise breakdown of total water resources (including surface water allocation from SSP and groundwater resources) reveals the highest availability in Vadodara district at 2.20 lakh ha-m, followed by Ahmedabad with 2.08 lakh ha-m, Anand with 1.87 lakh ha-m, Mahisagar with 1.55 lakh ha-m, Kheda with 1.46 lakh ha-m, Panchmahal with 1.16 lakh ha-m, Chhotaudepur with 0.85 lakh ha-m and Dahod with 0.53 lakh ha-m.

The highest total water demand from agriculture, horticulture, livestock, domestic and industry sectors is observed in Ahmedabad district at 4.86 lakh ha-m, followed by Kheda with 4.25 lakh ha-m, Anand with 4.18 lakh ha-m, Vadodara with 2.23 lakh ha-m, Panchmahal with 1.20 lakh ha-m, Mahisagar with 1.01 lakh ha-m and Chhotaudepur with 1.00 lakh ha-m.

Out of the eight districts in the Central Gujarat Zone, only Mahisagar district has surplus water resources, amounting to 0.54 lakh ha-m (53.66%). The other seven districts face a shortfall in water resources, with Kheda district experiencing the highest deficit of 65.70%, followed by Dahod with deficit of 62.48%, Ahmedabad with 57.15% deficit, Anand with 55.22% deficit and the lowest deficit was reported in Vadodara district (1.30%)..

Table 3: Water Budgeting for the Different Districts of the Central Gujarat Zone, Gujarat

District	*Total Surface and Ground Water Resources (ha-m)*	*Total Water Demands (ha-m)*	*Water Surplus (+)/Deficit (–) (ha-m)*	*Surplus (+)/*Deficit *(–) of Requirements (%)*
Ahmedabad	208435	486410	–277975	–57.15
Anand	187229	418147	–230918	–55.22
Chhotaudepur	85927	100683	–14756	–14.66
Dahod	53702	143126	–89424	–62.48
Kheda	146123	425966	–279843	–65.70
Mahisagar	155520	101211	54309	53.66
Panchmahal	116592	120010	–3419	–2.85
Vadodara.	220818	223728	–2911	–1.30
Total Central Gujarat	1174345	2019281	–844936	–41.84

In summary, the total water resources available in the Central Gujarat Zone are 11.74 lakh ha-m, while the total water demands amount to 20.1 lakh ha-m, resulting in a shortfall of –8.44 lakh ha-m. This represents a deficit of –41.84% compared to the total water demands.

South Gujarat Zone

The South Gujarat zone (Figure 10) encompasses a total geographical area of 22.28 lakh hectares, comprising eight districts viz; Bharuch (5.10 lakh ha), Dang (1.76 lakh ha), Narmada (2.82 lakh ha), Navsari (2.21 lakh ha), Surat (4.30 lakh ha), Tapi (3.14 lakh ha) and Valsad (2.95 lakh ha). Table 4 provides detailed water budgeting information for the different districts of the South Gujarat zone.

The district-wise breakdown of total water resources (including surface and groundwater resources) reveals the highest availability in Tapi district at 7.42 lakh ha-m, followed by Surat with 2.16 lakh ha-m, Bharuch with 1.87 lakh ha-m, Navsari with 1.19 lakh ha-m, Valsad with 1.13 lakh ha-m, Narmada with 0.98 lakh ha-m and Dang with 0.31 lakh ha-m.

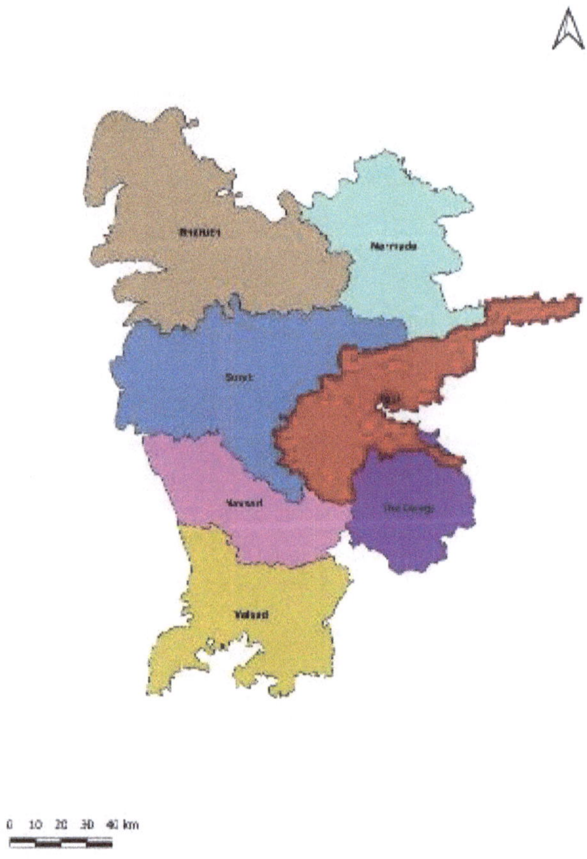

Figure 10: South Gujarat Zone

In the South Gujarat zone, the district-wise total water demands from agriculture, horticulture, livestock, domestic and industry sectors are as follows: Surat district has the highest demand at 2.53 lakh ha-m, followed by Bharuch with 1.96 lakh ha-m, Tapi with 1.46 lakh ha-m, Navsari with 1.14 lakh ha-m, Narmada with 0.52 lakh ha-m, Valsad with 0.42 lakh ha-m and Dang with 0.13 lakh ha-m.

Out of the seven districts in the South Gujarat Zone, five districts have surplus water resources, while two districts face a shortfall. Tapi district leads with the highest surplus water resources of 5.965 lakh ha-m (408.04%), followed by Valsad with 0.702 lakh ha-m (163.98%) and

Narmada with 0.46 lakh ha-m (88.09%). On the other hand, Surat district experiences the highest shortfall in water resources at –0.374 lakh ha-m (–14.74%), followed by Bharuch with –0.0899 lakh ha-m (–4.59%).

The total water resources available in the South Gujarat Zone amount to 15.08 lakh ha-m, whereas the total water demands are 8.19 lakh ha-m. These results in a surplus of 6.88 lakh ha-m in the South Gujarat Zone, representing a surplus of 84.05% compared to the total water demands.

Table 4: Water Budgeting for the Different Districts of the South Gujarat Zone, Gujarat State

District	*Total Surface and Ground Water Resources (ha-m)*	*Total Water Demands (ha-m)*	*Water Surplus (+)/ Deficit (–) (ha-m)*	*Surplus (+)/ Deficit (–) of Requirements (%)*
Bharuch	187042	196033	–8991	–4.59
Dang	31198	13289	17908	134.76
Narmada	98442	52339	46103	88.09
Navsari	119317	114859	4458	3.88
Surat	216489	253921	–37432	–14.74
Tapi	742638	146177	596462	408.04
Valsad	113048	42825	70224	163.98
Total South Gujarat	1508174	819442	688732	84.05

Saurashtra Zone

The Saurashtra zone (Figure 11) covers a total geographical area of 60.25 lakh hectares, comprising eleven districts: Amreli (7.39 lakh ha), Bhavnagar (6.58 lakh ha), Botad (2.63 lakh ha), Devbhoomi Dwarka (4.12 lakh ha), Gir Somnath (3.77 lakh ha), Jamnagar (6.61 lakh ha), Junagadh (5.05 lakh ha), Morbi (4.98 lakh ha), Porbandar (2.32 lakh ha), Rajkot (7.60 lakh ha) and Surendranagar (9.15 lakh ha). The water budgeting information for the different districts of the Saurashtra zone in Gujarat State is given in Table 5.

The district-wise breakdown of total water resources (including surface water allocation from SSP and groundwater resources) reveals the highest availability in Surendranagar district at 2.60 lakh ha-m, followed by Jamnagar with 1.86 lakh ha-m, Rajkot with 2.31 lakh ha-m, Bhavnagar with 1.78 lakh ha-m, Junagadh with 1.70 lakh ha-m, Amreli with 1.68 lakh ha-m, Morbi with 1.40 lakh ha-m, Gir Somnath with 0.87 lakh ha-m, Botad with 0.85 lakh ha-m, Devbhoomi Dwarka with 0.56 lakh ha-m and Porbandar with 0.40 lakh ha-m.

The highest total water demand from agriculture, horticulture, livestock, domestic and industry sectors is observed in Rajkot district at 2.53 lakh ha-m, followed by Surendranagar with 2.39 lakh ha-m, Junagadh with 1.44 lakh ha-m, Jamnagar with 1.42 lakh ha-m, Bhavnagar with 1.387 lakh ha-m, Amreli with 1.384 lakh ha-m, Morbi with 1.05 lakh ha-m, Botad with 0.98 lakh ha-m, Gir Somnath with 0.82 lakh ha-m, Devbhoomi Dwarka with 0.46 lakh ha-m and Porbandar with 0.43 lakh ha-m.

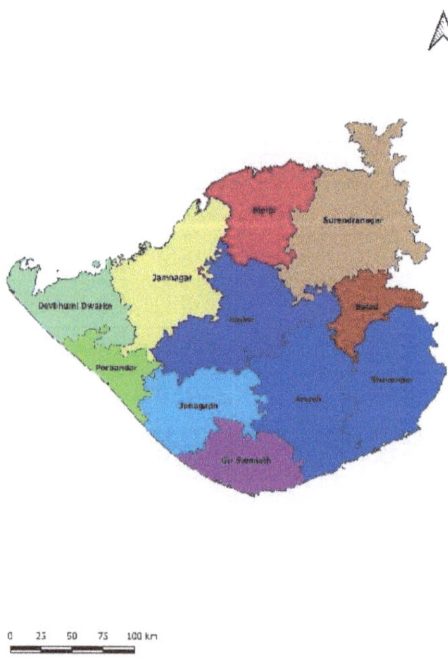

Figure 11: Saurashtra Zone

Out of the eleven districts in the Saurashtra Zone, eight districts have surplus water resources, while the other three districts face a shortfall. Morbi district leads with the highest surplus water resources of 0.35 lakh ha-m (33.02%), followed by Jamnagar with 0.442 lakh ha-m (31.01%) and Bhavnagar with 0.397 lakh ha-m (28.65%). On the other hand, Rajkot district experiences the highest shortfall in water resources at –0.22 Lakh ha-m (–8.71%), followed by Botad with –0.125 lakh ha-m (–12.73%) and Porbandar with –0.0335 lakh ha-m (–7.63%).

The total water resources available in the Saurashtra Zone amount to 16.07 lakh ha-m as against the total water demands of 14.33 lakh ha-m. This indicates a surplus of 1.73 lakh ha-m in the Saurashtra Zone, representing a surplus of 12.13% compared to the total water demands.

Table 5: Water Budgeting for the Different Districts of the Saurashtra Zone, Gujarat State

District	*Total Surface and Ground Water Resources (ha-m)*	*Total Water Demands (ha-m)*	*Water Surplus (+)/ Deficit (–) (ha-m)*	*Surplus (+)/Deficit (–) of Requirements (%)*
Amreli	168598	138456	30142	21.77
Bhavnagar	178470	138729	39742	28.65
Botad	85944	98477	–12533	–12.73
D.B.Dwarka	56397	46507	9890	21.26
Gir Somnath	87261	82066	5195	6.33
Jamnagar	186889	142652	44236	31.01
Junagadh	170851	144417	26434	18.30
Morbi	140861	105898	34963	33.02
Porbandar	40595	43949	–3354	–7.63
Rajkot	231276	253340	–22064	–8.71
Surendranagar	260738	239467	21271	8.88
Total Saurashtra	1607880	1433958	173923	12.13

Kachchh Zone

The Kachchh zone, encompassing a total geographical area of 43.16 lakh hectares (Figure 12), is summarized in Table 6, providing an overview of the total water supplies from various sources and the total demands from different sectors. The total water resources in the Kachchh zone, inclusive of surface water allocation from SSP and groundwater resources, amount to 2.02 lakh ha-m. Simultaneously, the total water demand from agriculture, horticulture, livestock, domestic and industry sectors in the Kachchh zone is estimated to be 1.93 lakh ha-m. Consequently, there is a surplus of 0.084 lakh ha-m of water resources in the Kachchh zone, representing a surplus of 4.39% over water demands.

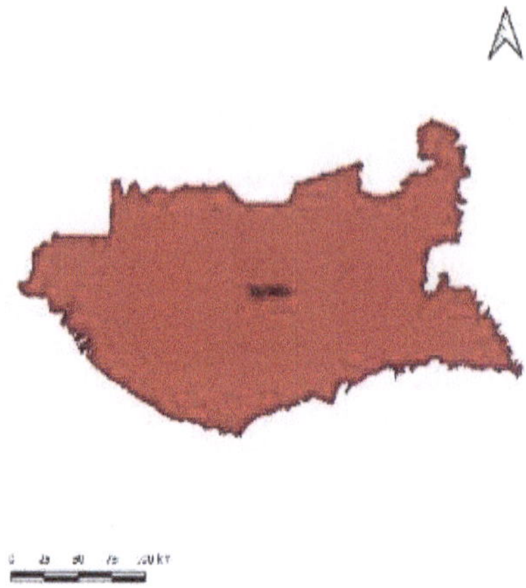

Figure 12: Kachchh Zone

Table 6: Water Budgeting for the Different Districts of the Kachchh Zone, Gujarat State

District	*Total Surface and Ground Water Resources (ha-m)*	*Total Water Demands (ha-m)*	*Water Surplus (+)/ Deficit (–) (ha-m)*	*Surplus (+)/Deficit (–) of Requirements (%)*
Kachchh	201928	193433	8494	4.39

Gujarat State

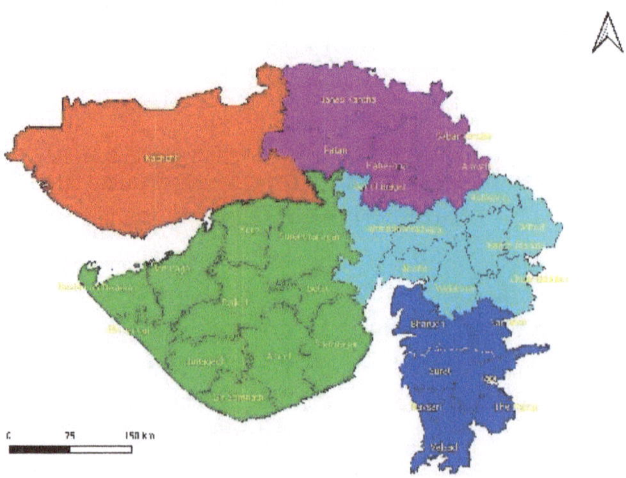

Figure 13: Zones of Gujarat State

he state of Gujarat is divided into five zones, namely North Gujarat (30.443 lakh ha), Central Gujarat (30.095 lakh ha), South Gujarat (22.281 lakh ha), Saurashtra (60.25 lakh ha) and Kachchh (43.16 lakh ha) (Figure 13), each characterized by its distinct geographical area.

Water Resources Availability

Table 7 presents the availability of surface water resources, including allocation from SSP (Sardar Sarovar Project) and groundwater resources in different zones of Gujarat state. The total water resource availability in Gujarat state is 53.73 lakh ha-m, comprising groundwater resources of 24.58 lakh ha-m, surface water allocation of 9.15 lakh ha-m from SSP and of 19.99 lakh ha-m from major, medium and minor projects other than SSP. In Gujarat State, the distribution of water resources shows that groundwater constitutes 45.75%, while surface water resources make up 54.25% of the total water resource.

Table 7: Surface and Groundwater Resources in Different Zones of Gujarat State

Sr No.	*Zone of Gujarat State*	*Total Geographical Area (ha)*	*Groundwater Resources (ha-m)*	*Surface Water Resources Other Than SSP (ha-m)*	*Surface Water Resources from SSP (ha-m)*	*Total Water Resource (ha-m)*
1.	North Gujarat	3044287	231999	425422	231999	22350
2.	Central Gujarat	3009541	3825600	476520	407178	290648
3.	South Gujarat	2228051	462384	953410	92381	1508174
4.	Saurashtra	6025004	1010822	344254	252804	1607880
5.	Kachchh	4316260	82961	62578	56389	201928
	Total Gujarat	18623143	2458108	1999418	915722	5373248

Water Demands

Table 8 provides an overview of the water demands across various sectors in different zones of Gujarat state. The agricultural cropped area in Gujarat spans 99.5345 lakh ha, with irrigation water requirements for agriculture crops estimated at 42.45 lakh ha-m. Additionally, the horticultural sector, covering all fruit, vegetable, flower and spice crops, encompasses 19.16 lakh ha, with irrigation water demands amounting to 7.07 lakh ha-m. The livestock population in Gujarat is recorded at 499.49 lakh, requiring 0.82 lakh ha-m of water. The total domestic water requirements for the state, catering to the human population of 716 lakh, are estimated at 2.779 lakh ha-m.

Water Budgeting

Table 9 presents the total water supplies from various sources and the total demands from various sectors for different zones in Gujarat state. The total water resource availability in the state is 53.73 lakh ha-m, while the total water demands amount to 54.76 lakh ha-m, resulting in a deficit of 1.02 lakh ha-m in Gujarat State. This deficit represents a 1.88% shortfall compared to the total water demands.

Table 8: Water Requirements from the Various Sectors in Different Zones of Gujarat State

Zone of the Gujarat State	*Agriculture*		*Horticulture*		*Livestock*		*Domestic*		*Industry*	*Total Water Demands (ha-m)*
	Cropped Area (ha)	*IWR (ha-m)*	*Cropped Area*	*IWR (ha-m)*	*Nos.*	*WR (ha-m)*	*Nos.*	*WR (ha-m)*	*WR (ha-m)*	
North Gujarat	2097705	763916	430653	143924	10169197	25999	13060241	45763	30297	1009899
Central Gujarat	2250558	1683379	360365	160931	19235036	24338	25700646	90055	60578	2019281
South Gujarat	875190.3	623065	288517	112763	7917186	8856	14319237	50175	24583	819442
Saura-shtra	4311606	1052852	708617	236099	10247977	18593	23799863	83395	43019	1433958
Kachchh-chh	418387	121666	127640	52814	2379386	4611	2437194	8540	5803	193433
Gujarat State Total	9953446	4244878	1915792	706531	49948782	82397	79317181	277928	164280	5476013

Table 9: Water Budgeting for the Different Zones of Gujarat State

Zone of Gujarat	*Total Water Resources- Surface + Ground Water (ha-m)*	*Total Water Demands (ha-m)*	*Water Surplus (+)/Deficit (–) (ha-m)*	*Surplus (+) /Deficit (–) of Requirements (%)*
North Gujarat	880921	1009899	–128979	–12.77
Central Gujarat	1174345	2019281	–844936	–41.84
South Gujarat	1508174	819442	688732	84.05
Saurashtra	1607880	1433958	173923	12.13
Kachchh	201928	193433	8494	4.39
Gujarat State	5373248	5476014	–102766	–1.88

Food Securities

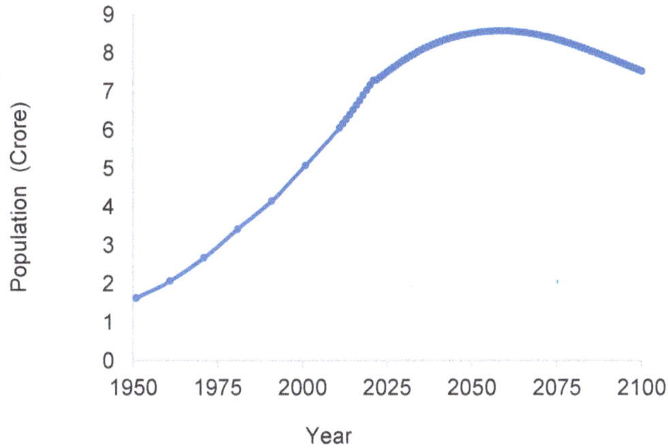

Figure 14: Ever-increasing Population of Gujarat

Agricultural production data were sourced from the Directorate of Agriculture, Department of Farmers Welfare and Cooperation, Gandhinagar (DoA, 2024). The food requirements for various components were estimated based on the current population of the Gujarat state, considering recommendations from the Indian Council of Medical Research (ICMR) for human nutrition (Table 10).

The estimation of agricultural produce requirements considered the present population and nutritional needs based on ICMR recommendations. The net sown area and gross cropped area are 103 and 131 lakh hectares, respectively, with net and gross irrigated areas covering 39 and 74.60 lakh hectares.

The food demand in Gujarat has risen in tandem with the increase in population, which has grown from 1.63 crores in 1951 to 7.17 crores as of January 1, 2024. Projections suggest a peak population of 8.56 crores by 2060, followed by a decrease to 7.52 crores in 2100 (Figure 14).

Table 10 reveals that, against the requirements of 94.1 lakh MT of cereals/millets and 31.4 lakh MT of pulses for the current population of 7.16 crores, there is a shortfall of production amounting to 13.02 lakh MT of cereals/millets and pulses by 4.40 lakh MT. Notably, there is a surplus in the production of 62.79 lakh MT of vegetables, 43.48 lakh MT of fruits, 88.82 lakh MT of milk, 37.16 lakh MT of fats and oils, 54.46 lakh MT of oilseeds and nuts, 8.34 lakh MT of spices, 1.62 lakh MT of flowers, 2.32 lakh MT of sugar and 47.77 lakh MT of cotton.

Table 10: Productions and Demands of Agricultural Produce for Food in Gujarat (2023)

Food Composition	*Requirements As per ICMR (g/day/person)*	*Requirements (Lakh MT)*	*Production (Lakh MT/Y)*	*Shortfall (–) or Surplus (+) (Lakh MT/Y)*
Cereals & Millets	360	94.1	81.06	–13.02
Pulses (Legumes)/ Fresh food	120	31.4	26.96	–4.40
Green Leafy vegetables	100	26.1	167.33	62.79
Other vegetables	200	52.3		
Roots and Tubers	100	26.1		
Sub Total		**104.5**		
Fruits	150	39.2	82.68	43.48
Milk	300	78.4	167.22	88.82
Fats and Oils	30	7.8	45.0	37.16
Oilseeds and nuts	30	7.8	62.3	54.46
Spices	10.49	2.7	11.09	8.34
Flowers	1.26	0.33	1.95	1.62
Sugar	30	7.84	10.16	2.32
Sugarcane		–	147	147.00
Cotton	24.66	6.44	54.21	47.77

Issues

The state of Gujarat exhibits a surplus in the production of most food items surpassing local consumption. Overall, Gujarat stands as a surplus state in terms of food production. However, specific shortfalls are observed, particularly in cereals/millets and pulses, signalling the necessity to augment the cultivation area for those crops. The following issues exist to harness the enhanced agricultural productions in the state.

- The rivers in Saurashtra, North Gujarat and Kachchh flow only during the monsoon season with limited water flow.
- There is inefficient irrigation infrastructure and the performance of existing water resource projects is low.
- There are limited opportunities to enhance surface water resources in Saurashtra, North Gujarat and Kachchh zones due to hydrological and topographical aspects.
- There is uneven distribution of rainfall and water resources in the different zones of Gujarat.
- The topography and economy hinder the transfer of water from surplus zones to deficit zones within the state.
- There are limited opportunities to enhance groundwater resources in a larger part of the state due to aquifer characteristics and the spatial and temporal variability of rainfall, which is often erratic.
- Over withdrawals of groundwater in the coastal belt area have accelerated sea water intrusions, deteriorating groundwater and soil quality.
- Excessive pumping of groundwater has led to a deep water table, poor water quality and high demands for electricity.
- Soils are being degraded due to irrigation with inferior quality water.
- Overapplication of irrigation water in South Gujarat and parts of Central Gujarat has resulted in waterlogged and saline soils.

More than half of the cultivated land in the state relies on rainfed farming, leading to low productivity of rainfed crops. Gujarat receives most of its rainfall during the monsoon, which is often erratic and inadequate, leading to droughts and floods that affect crop yields.

- Farmers are reluctant to sow cereal and pulse crops due to lower economic returns.
- Farmers tend to favour excessive use of fertilizers, weedicides, pesticides, etc., over farmyard manure (FYM) due to labour shortages, leading to environmental and soil pollution.
- Monoculture cropping depletes soil nutrients and increases pest susceptibility.
- Limited access to and awareness of advanced agricultural technologies hinders the adoption of modern irrigation and cultivation practices, affecting productivity and resource efficiency.
- Farmers often face volatile market prices, making it difficult to earn a stable income. Small and marginal farmers have limited negotiating power with intermediaries and market forces. Lack of adequate storage facilities and access to markets leads to post-harvest losses.

- Many farmers rely on loans due to low income and unpredictable harvests, leading to debt burdens. Agriculture-dependent families have limited alternatives during off-seasons or crop failures, increasing vulnerability. Issues like illiteracy and lack of social security further exacerbate farmer grievances.
- Climate change has severely affected agricultural production.

Policy Suggestions

- The performance of irrigation projects should be enhanced to their potential by strengthening water distribution networks, either through canal lining or adopting Pressurized Irrigation Network Systems (PINs).
- Plenty of water resources are available in South Gujarat zone, which can be transferred to the northern part of Central Gujarat, North Gujarat, Kachchh and Saurashtra zones. Investment in such infrastructure should be given top priority.
- A channel linking all rivers draining into the coastal belt of the state should be excavated at a feasible distance of 40–50 km upstream of the seashore to prevent sea water intrusion and enhance groundwater resources and quality.
- Promote in-situ water harvesting within fields through farming practices, contour bunding, ridge and furrow systems, etc.
- Adopt water harvesting-cum-groundwater recharge structures in waste/fallow land as well as on streams in series.
- Promote the adoption of water-efficient irrigation methods along with new dimensions of water management strategies, such as mulching, fertigation, subsurface irrigation, pulsed drip irrigation, deficit irrigation, aerated irrigation, etc.
- In response to changing climate, adopt crop varieties that can sustain extremities of weather and rainfall.
- Promote the reclamation of problematic soils to enhance agricultural production in the state.
- Encourage organic farming through subsidies and training to farmers to promote soil health and reduce pollution. Provide organic farmers with separate marketplaces to harness higher selling prices.
- Promote crop diversification through crop rotation and intercropping to improve soil fertility and pest management.
- Expand and strengthen the extension services to increase the access to agricultural knowledge and modern technologies through training programs and field demonstrations.

- Strengthen farmer cooperatives/FPOs to improve bargaining power and access to markets. Improve market infrastructure through adequate storage facilities and cold chains to reduce post-harvest losses. Implement fair pricing mechanisms through MSP schemes and direct market access platforms to ensure fair returns for farmers.
- The farmer distress should be reduced by restructuring options to ease the burden on indebted farmers. The livelihood diversification should be promoted by enhancing the off-farm income opportunities through decentralized agro industries, rural entrepreneurships and skill development programs.
- The social safety nets for farmers should be strengthened by expanding access to healthcare, insurance and other social security programs.
- The infrastructures for the industries, commercial and recreational activities should be prioritized in wasteland or problematic soils.

Summary

North Gujarat Zone

In North Gujarat, the total water resources encompass 4.25 lakh ha-m from groundwater, 2.32 lakh ha-m from surface water (excluding SSP) and 2.24 ha-m from SSP allocation. The distribution shows that 48.3% is from groundwater and 51.7% is from surface water. However, the total water availabilities and demands in North Gujarat are 8.81 lakh ha-m and 10.1 lakh ha-m, resulting in a shortfall of –1.29 lakh ha-m. This signifies a deficit of –12.77% in water resources compared to the total water demands in the zone.

Central Gujarat Zone

In Central Gujarat, the total water resources comprise 4.77 lakh ha-m from groundwater, 4.07 lakh ha-m from surface water (excluding SSP) and 2.91 lakh ha-m from SSP allocation. The distribution indicates that 40.57% is from groundwater, while 59.43% is from surface water. However, the total water resources availabilities and demands are 11.74 lakh ha-m and 20.1 lakh ha-m, respectively, resulting in a shortfall of –8.44 lakh ha-m. This deficit represents –41.84% in comparison to the total water demands in the zone.

South Gujarat Zone

In South Gujarat, the total water resources include 4.62 lakh ha-m from groundwater, 9.53 lakh ha-m from surface water (excluding SSP) and

9.24 lakh ha-m from SSP allocation. The distribution indicates that 30.65% is from groundwater, while 69.34% is from surface water. The total water resources availabilities and water demands are 15.08 lakh ha-m and 8.19 lakh ha-m, respectively, resulting in a surplus of 6.88 lakh ha-m. This surplus represents 84.05% more than the total water demands in the zone.

Saurashtra Zone

In Saurashtra, the total water resources include 10.11 lakh ha-m from groundwater, 3.44 lakh ha-m from surface water (excluding SSP) and 2.53 lakh ha-m from SSP allocation. The distribution indicates that 62.86% is from groundwater, while 37.13% is from surface water. The total water resources availabilities and water demands are 16.07 lakh ha-m and 14.33 lakh ha-m, respectively, resulting in a surplus of 1.73 lakh ha-m. This surplus represents a 12.13% excess compared to the total water demands in the Zone.

Kachchh Zone

In Kachchh, the total water resources include 0.83 lakh ha-m from groundwater, 0.63 lakh ha-m from surface water (excluding SSP) and 0.56 lakh ha-m from SSP allocation. The distribution indicates that 41.08% is from groundwater, while 58.9% is from surface water. The total water resources availabilities and water demands are 2.01 lakh ha-m and 1.93 lakh ha-m, respectively, resulting in a surplus of 0.085 lakh ha-m in the Kachchh Zone. This surplus represents a 4.39% excess compared to the total water demands.

Gujarat State

In Gujarat, the state's total water availability stands at 53.73 lakh ha-m, slightly falling short of the total water demands of 54.76 lakh ha-m, resulting in a deficit of 1.02 lakh ha-m or a –1.88% shortfall. In context to the food security landscape, the current population's needs include 94.1 lakh MT of cereals/millets and 31.4 lakh MT of pulses. However, there is a production shortfall of –13.02 lakh MT of cereals/millets and –4.4 lakh MT of pulses. However, there is a surplus in the production of various agricultural products, including 62.79 lakh MT of vegetables, 43.48 lakh MT of fruits, 88.82 lakh MT of milk, 37.16 lakh MT of fats and oils, 54.46 lakh MT of oilseeds and nuts, 8.34 lakh MT of spices, 1.62 lakh MT of flowers, 2.32 lakh MT of sugar and 47.77 lakh MT of cotton. Balancing these disparities requires a strategic and integrated approach to water management and agricultural practices, ensuring a more sustainable and secure future for Gujarat's water and food resources.

Despite the current relatively small deficit of 1.88%, prudent water utilization is crucial. Future scenarios, shaped by climate change and increasing demands for food, fibre, fruits and vegetables to sustain a growing population and rising living standards, are expected to lead to increased water demands. Consequently, proactive measures are required in Gujarat State, such as enhancing water resources through *in-situ* water harvesting and artificial groundwater recharge. The adoption of water-efficient irrigation methods, including Micro Irrigation Systems (MIS), mulching and deficit irrigation applications, is essential. Encouraging the cultivation of low-water-demanding crops like pulses, spices and millets is advisable. Government initiatives should promote the adoption of MIS, mulching and deficit water applications among farmers to expand the irrigated area.

References

[1] DoA (2024). Directorate of Agriculture, Department Agriculture, Farmers Welfare and Cooperation, GoG, https://dag.gujarat.gov.in/, visited on 11.01.2024.

[2] GGRC (2024). Gujarat Green Revolution Limited, Vadodara. www.ggrc.co.in visited on 15.01.2024.

[3] GWRD (2024). https://guj-nwrws.gujarat.gov.in/ visited, 25.12.2023. Gujarat Water Resources Department. GoG, Gandhinagar

[4] MoWR, GoG (2024). Ministry of Water resources, Government of Gujarat, 1st All India Annual States' Ministers Conference on "Water Vision@2047" of the water resources ministers of 30 states, Union ministers and officials., at Bhopal, MP, India., The Times of India, January 3, 2024. https://timesofindia.indiatimes.com/city/ahmedabad/water-availability-to-double-in-25-years/articleshow/96779461.cms

[5] The Wire (2024). https://thewire.in/agriculture/govt-mismanagement-is-coming-in-the-way-of-irrigation-water-for-gujarats-farmers, visited on 04.01.2024.

Water Security for Agriculture—*Atmanirbhar* Farmers

A.R. Pathak[1] and P.A. Pandya[1]

ABSTRACT: India, with its significant agricultural sector and diverse water challenges, has been exploring various technological interventions to enhance water security in agriculture. Several technologies have been implemented to improve water efficiency, reduce wastage and ensure sustainable use of water resources. The article presents the importance of water security for making Indian Agriculture and Farmers Atmanirbhar. The article highlights the critical connection between agriculture, water security and global concerns regarding water scarcity. In India, water security faces challenges such as over-extraction of groundwater, pollution and unequal distribution. Agriculture, being the largest consumer of water globally, necessitates innovative solutions. Technological interventions, including Micro Irrigation Systems (MIS), water harvesting and smart irrigation, can enhance water efficiency. Remote sensing and GIS technologies offer valuable tools for monitoring and managing water resources. Mobile applications aid farmers with real-time information, while alternative cropping systems and biotechnological interventions contribute to water-efficient practices. Soil health management, irrigation water quality, deficit irrigation and mulching technologies further address sustainable water use. The paper emphasizes the multifaceted nature of water security, especially in the context of agriculture and advocates for a holistic and collaborative approach involving government, civil society, private sector and communities to ensure sustainable water management and overall economic and social development. To make more meaningful and efficient contributions to addressing water security and resilience across various sectors and scales, there is a need for interdisciplinary research and the establishment of productive synergies between policymakers and researchers.

Keywords: *Atmanirbhar (Self-reliant) Agriculture, Water security, Micro Irrigation Systems (MIS), Precision Agriculture, Sustainable water management, Technological interventions, Water harvesting.*

---

1 Former Vice Chancellor, Junagadh Agricultural University, Junagadh & Navsari Agricultural University, Navsari, Gujarat.

Introduction

Agriculture, serving as the backbone of India's economic structure, plays a crucial role in the country's development. It is a significant source of employment, providing livelihoods to a large percentage of the population, particularly in rural areas. The sector's contribution to the overall economic output has ripple effects on various industries. Ensuring food security, supplying raw materials for industries and earning foreign exchange through exports make agriculture indispensable to the Indian economy. Recognizing these facts, achieving self-reliance (*Atmanirbharta*) in India necessitates a focus on *Atmanirbhar* Agriculture and *Atmanirbhar* Farmers. The key challenges to making the Agriculture and Farmers *Atmanirbhar in India is displayed in Figure 1.*

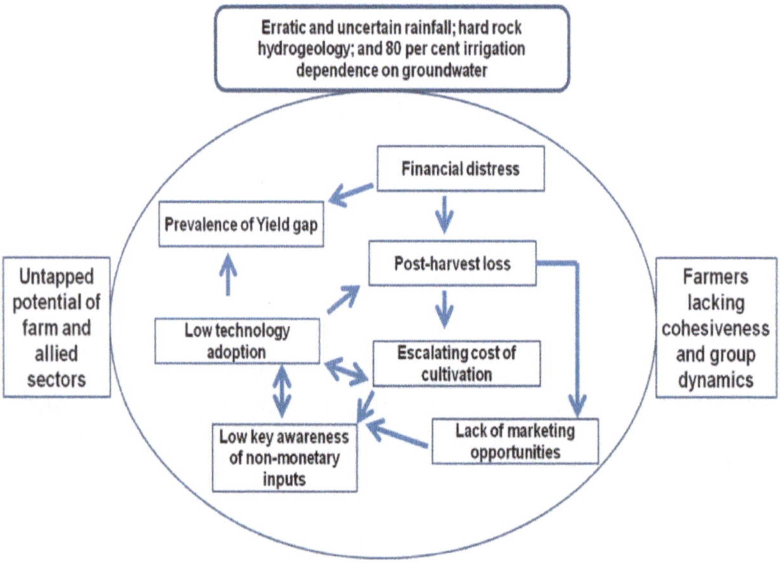

Figure 1: Key Challenges Against to Make the Farmers *Atmanirbhar*

Water Security and World

Water security refers to the availability of adequate and clean water resources to meet the country's growing water demand, while ensuring the sustainability of the water ecosystems. This is in line with the definition of water security provided by UN Water, which is a United Nations inter-agency coordination mechanism on all freshwater-related issues. Achieving water security is a pressing global concern and stands out as one of the foremost challenges in the 21st century, as highlighted by various researchers (Scott *et al.*, 2021). With only 2.5% of the world's total water resources available as freshwater for human use and global water consumption increasing by 1% annually, the urgency of addressing water security becomes apparent (Castro-Pardo *et al.*, 2022).

Given that the agricultural sector stands as the largest consumer of water globally, water security assumes particular importance in this sector (FAO, 2017). Projections indicate that by 2035, the agricultural sector's share in total water extraction from global freshwater resources will reach 70%, exceeding 90% in less developed countries (Eidi *et al.*, 2020). Agricultural water security extends beyond water availability and includes considerations of food security, environmental security, agricultural economic security and rural community security (Valizadeh *et al.*, 2021). Consequently, a failure to ensure water security has profound and irreversible impacts on raw material production, rural employment opportunities and rural development (Taylor, 2015).

Globally, agriculture constitutes the largest share (72%) of total annual freshwater withdrawals, with significant variations across regions and countries (World Bank 2019a). South Asia, according to the World Bank, leads in agricultural water use (91%), while the European Union trails with the smallest share (31%). India, among the top agricultural producers globally, tops the list in agricultural water use (90%), followed by the People's Republic of China (PRC) (64%), Brazil (61%) and the United States (40%) (World Bank 2019b). Moreover, water use in the agriculture sector in least developed countries accounts for about 91% of the global total (Table 1). Despite increasing water demand from non-agricultural sectors, agriculture is projected to maintain significantly higher absolute water demand for the next two decades (Gruère and Shigemitsu, 2021). This stems from the imperative to meet the food demand of a rising global population, expected to reach around 10 billion people in 2050 (United Nations 2022).

With higher economic growth in developing countries, per capita food demand growth is expected to outpace population growth alone, emphasizing the need for a self-reliant farming approach and supporting the well-being of farmers. To address the challenges ahead, it is crucial to consider the implications of climate change on water resource availability and spatial distributions in the global food supply.

Table 1: Share of Agriculture in Total Annual Freshwater Withdrawals, 2019

Country/Region	*Freshwater Used in Agriculture (%)*	*Renewable Internal Freshwater Resources, Total (billion m^3)*	*Level of Water Stressa*
Brazil	61	5,661	1.55
People's Republic of China	64	2,813	43.22
India	90	1,446	66.19
United States	40	2,881	28.16
World	72	42,809	
Least developed countries	91	-	
South Asia	91	-	
East Asia and the Pacific	73	-	

[a]Freshwater withdrawals as a proportion of available freshwater resources.
Source: Food and Agriculture Organization (FAO), AQUASTAT data.

Water Security and India

India faces a formidable challenge in supporting a substantial portion of the global human and livestock population, accounting for 18% and 15%, respectively, with only 2.4% of the world's land and 4% of its water resources. This imbalance poses a significant challenge, aggravated by the impact of climate change, substantial wastage due to poor management practices and distorted water pricing policies. The country's water resources exhibit notable variations in availability and demand across different regions, indicating inefficient and inequitable utilization and distribution practices. The total utilizable water resources are estimated at 1123 billion cubic meters (BCM), comprising 690 BCM from surface water and 433 BCM from groundwater sources. As India grapples with an ever-growing population and an increase in living standards, the demand for water resources is expected to surge to meet requirements in food production, domestic use and industrial activities. Projections indicate a total water demand of 1447 BCM by 2050, surpassing the current level of utilizable water resources. Agriculture alone is anticipated to account for 1074 BCM of this demand (Pathak *et al.*, 2021). Addressing the water security against this looming water crisis requires a holistic and sustainable approach, encompassing efficient water management, equitable distribution and policies that align with the evolving needs of the nation.

Technological Interventions for Achieving Water Security in Agriculture

Micro Irrigation System

Micro irrigation systems represent an efficient approach for water security in agriculture as it delivers water directly to the plant root zone, minimizing water wastage and optimizing crop yield. Research endeavours focusing on energy conservation, water efficiency and economic considerations for various crops in Gujarat, as outlined in Table 2, underscore the importance of continuous improvement in micro irrigation technologies.

Table 2: Results of MIS Research by SAUs of Gujarat on Various Crops

Crop	*Total Area Under MIS (ha)*	*Total Saving of Labour Cost in Irrigation Application, Weedings and Fertilizer (Rs. lakh)*	*Saving in Energy Use (Rs. lakh)*	*Additional Income from Increased in Yield (Rs. lakh)*	*Cost Saving and Additional Income from Increased Production (Rs. lakh)*	*Additional Income from Water Saving (Rs. lakh)*	*Total Benefits Due to MIS Adoptions (Rs. lakh)*
(1)	*(2)*	*(3)*	*(4)*	*(5)*	*(6) = (3)+(4)+(5)*	*(7)*	*(8)=(6)+(7)*
Potato	3081	158	197	2193	2548	658	3206
Groundnut	609	28	19	579	626	509	1135
Guava	580	35	5	220	260	142	419
Lemon	4892	499	21	533	1053	265	2140
Coconut	743	68	12	47	127	41	298
Bt cotton	370449	17230	4932	54975	77137	36650	113787
Sugarcane	8486	805	258	3055	4118	2304	6422
Brinjal	1974	104	43	1283	1430	1055	2485
Banana	18330	4859	416	21996	27270	13064	40335
Castor	6280	344	56	678	1078	430	1508
Bitter Guard	768	57	12	292	360	197	557
Papaya	5128	660	131	3333	4124	2222	6346

Investing in R&D initiatives is essential to elevate the overall efficiency and cost-effectiveness of micro irrigation, necessitating the development of custom-made components, materials and systems tailored to the unique requirements of diverse regions across India. Pandya and Dwivedi

(2016) highlighted the major constraints faced by the farmers in adoption of drip irrigation such as high Initial Investment, damage to system due to rats and other animals, non-availability of technical services at village level for maintenance and repair of MIS systems, clogging of system due to salty/impure water and not up to the mark after sales service needs to be addressed by research and policy interventions to upscale the adoption of micro irrigation. To mitigate potential damage from stray animals, it is recommended to incorporate chemical substances into the raw materials used for manufacturing PVC/HDPE/LLDPE components of MIS. Establishing MIS operation and maintenance clinics at the taluka level can effectively address these challenges. Moreover, deploying technical experts in irrigation water management at the taluka level can contribute to the successful implementation and upkeep of micro irrigation systems. A robust monitoring and evaluation system needs to be established to track the performance and impact of micro irrigation systems. This information can be used to refine strategies, address challenges and improve the adoption process.

Water Harvesting and Aquifer Recharge

Implementing advanced water harvesting and artificial groundwater recharge is crucial for ailing water security in arid and semiarid regions. Strategic basin-scale water harvesting measures, including *in-situ* soil moisture conservation, can make drought-prone areas less vulnerable. Transferring surface water, constructing farm ponds and promoting less water-demanding development contribute to comprehensive water management. Small check dams strategically spaced along rivers increase groundwater recharge time is pivotal for minimizing evaporation losses and pumping costs in irrigation. Decentralized water harvesting-cum-groundwater recharge structures in reservoir catchment areas deserve priority. Technology-facilitated community-based water management, incentivizing cooperatives for groundwater recharge and solar-powered irrigation can foster water security in water-scarce regions. The research focus on how to select the sites for structure construction, the types of structure and their designs depending upon the local hydrogeology, groundwater flow regime, terrain condition and demand of water and how they impact on resource rejuvenation and improvement in water quality is essential for optimum water harvesting plan (Saha *et al.*, 2022).

Precision Agriculture with Smart Irrigation Systems

The adoption of Micro Irrigation Systems (MIS) alone is insufficient. The efficient and regular operation of the system, coupled with the application of water at the crop's required rate, is essential. Human interventions and the lack of technical knowledge among farmers often lead to water wastage by applying more water than necessary. Addressing these challenges requires the implementation of AI and IoT-based automatic irrigation systems. Presently, less than 1% of the total area under MIS has adopted automatic irrigation systems due to technical expertise limitations and higher initial costs, which may hinder the water security in agriculture (Boursianis *et al.*, 2020).

Precision farming, enabled by technologies such as sensors, GPS and data analytics, empowers farmers to optimize water and resource utilization through real-time insights into soil conditions, crop health and irrigation needs. Automated irrigation systems, equipped with sensors and controllers, assist farmers in efficiently managing water usage. The installation of soil moisture sensors in fields allows farmers to monitor soil moisture levels, facilitating precise irrigation management and preventing both over-irrigation and under-irrigation. Research efforts should prioritize the development of low-cost and user-friendly smart/automatic irrigation controllers to facilitate broader adoption and enhance sustainable agricultural practices.

Remote Sensing and GIS Technology

Utilizing satellite data and remote sensing technologies is crucial for monitoring crop health, identifying water stress and optimizing irrigation scheduling. This data empowers farmers to make informed decisions about water use. Remote sensing satellites provide high-resolution imagery, enabling farmers to monitor crop health, detect areas of stress, disease, or pest infestation and implement timely interventions. GIS technology integrates spatial data on soil moisture, weather patterns and crop types, allowing for the calculation of water use efficiency and optimizing irrigation practices. Remote sensing aids in early detection and monitoring of drought conditions, generating drought risk maps through vegetation indices and thermal data analysed by GIS. Remote sensing and GIS technology have proven to be effective tools in monitoring and early warning of agricultural droughts (Pandya and Gontia, 2023). GIS-based decision support systems integrate various information layers, such as climate data, soil characteristics and crop requirements, assisting farmers and water resource managers in making informed decisions and helps in water security of the region. It can serve as valuable tools for monitoring changes in land cover, land use and water availability over time, offering a historical perspective for understanding long-term trends and planning future water management strategies.

Mobile Applications and Advisory Services

Mobile applications and advisory services play a crucial role in modern agriculture, providing farmers with real-time information, insights and support for various aspects of farming. Mobile applications and advisory services provide farmers with information on weather forecasts, crop management practices and water conservation techniques which is critical for addressing water security. These digital platforms empower farmers to make informed decisions about their irrigation practices. Bhagat *et al.* (2023) evaluated farmers' perception of mobile application-based climate-resilient AgroMet Advisory Service and stressed upon the development of strong institutional mechanisms for addressing knowledge gaps related to climate-resilient agriculture and ensuring the successful implementation of region-specific climate-resilient agriculture action plans.

Alternate Cropping Systems and Biotechnology

Alternative cropping systems and optimized management practices has proven useful in saving groundwater and enhancing economic and environmental sustainability (Rathore *et al.*, 2022). Implementing alternative cropping systems is a key strategy for enhancing water security in agriculture. These systems involve cultivating crops that are well-suited to local environmental conditions and require less water. Diversifying crop choices by introducing drought-resistant or water-efficient varieties helps optimize water usage. Additionally, incorporating agroforestry practices, where trees or shrubs are integrated with crops, can contribute to improved water retention in the soil. Crop rotation, cover cropping and intercropping are alternative systems that enhance water efficiency by reducing evaporation and promoting soil moisture conservation. Agroecological approaches, such as permaculture and organic farming, prioritize sustainable water management by fostering natural processes that enhance soil structure and water retention. Integrating these alternative cropping systems into agricultural practices not only conserves water but also promotes resilience in the face of changing climatic conditions, contributing to long-term water security.

Adopting drought-resistant and water-efficient crop varieties through biotechnological interventions represents a transformative approach to addressing the challenges of water scarcity in agriculture. Biotechnology enables the development of crop varieties that exhibit enhanced tolerance to drought conditions and demonstrate improved water-use efficiency. These genetically modified crops possess traits that allow them to thrive in water-stressed environments, requiring less irrigation compared to traditional varieties. By harnessing the power of biotechnology, scientists can introduce genes responsible for traits such as improved water retention, reduced transpiration rates and efficient nutrient utilization. The adoption of these biotechnologically engineered crops not only mitigates the impact of water scarcity on agriculture but also contributes to sustainable and resilient farming practices, ensuring food security in regions prone to water-related challenges.

Soil Health Management

The adoption of integrated sustainable practices for soil and water management, including minimum tillage, crop rotation, organic matter addition and cover cropping, is vital for enhancing soil health, minimizing erosion and improving water infiltration and storage. These practices not only contribute to the preservation of soil biodiversity, enhanced fertility and carbon sequestration but also play a pivotal role in mitigating the impacts of climate change and bolstering food and nutrient security (Das *et al.*, 2022). Ensuring soil health is imperative for successful crop production and water security. Even with optimal inputs and management, including irrigation water applications, degraded land tends to yield lower outputs. In regions prone to soil salinity, agricultural land should be strategically flooded through the creation of field boundary bunds during the monsoon season. This practice, implemented for the crop tolerance period, effectively reduces soil salinity. Furthermore, for waterlogged areas, the implementation of location-specific designed drainage systems is essential to manage excess water and prevent waterlogging issues. Improving the water productivities of alkali soils can be achieved through the judicious addition of soil amendments and the promotion of natural farming practices. Reclaiming and managing these lands require strengthening watershed development and management programs. By implementing these measures, agricultural lands can be restored, ensuring enhanced soil health and water security.

Managing Quality of Irrigation Water

Escalating water demands across various sectors have led to the excessive withdrawal of groundwater, causing a concerning decline in the groundwater table. Water extracted from deeper depths is of poor quality and consumes more energy, posing challenges against water. The quality of irrigation water significantly influences crop yield and water productivity. Urbanization, industrialization and extensive chemical use in agriculture have contributed to water resource pollution. Enforcing conjunctive use of surface and groundwater in irrigation project command areas is vital for optimal water management. Strengthening pollution control boards is essential for monitoring and enforcing safe effluent disposal, encouraging reuse or recharge. Establishing efficient conveyance systems and treatment plants for urban sewage water prevents water body contamination. Coastal belts, vulnerable to seawater intrusion, require immediate attention, including constructing interceptor channels parallel to the coast to divert river flow and create a freshwater head above sea level. In water-scarce regions, exploring desalination technologies is imperative. Converting saline water into freshwater for agriculture provides an additional sustainable water source. These measures collectively contribute to responsible water management practices for a more sustainable future. Demerdash et *al.* (2022) developed quality-based irrigation water security index and reported that expanding drainage and sewage water treatment improved the developed index to high security levels.

Deficit Irrigation

Deficit irrigation, also known as regulated deficit irrigation, represents a strategy aimed at water saving by achieving higher yields per unit of applied irrigation water. This approach involves subjecting crops to a controlled level of water stress, either during specific growth periods or throughout the entire growing season. Empirical findings have demonstrated the suitability of deficit irrigation practices for various crops, including cotton, maize, wheat, sunflower, sugar beet and potatoes, where reduced evapotranspiration is applied consistently across the growing season. Also, the DI strategies offer great opportunities for saving water without compromising production. Several researchers have reported water savings from 43% to 65% under RDI strategy with a small reduction in yield, but with higher quality of produce (Mirás-Avalos *et al.*, 2016). Accordingly, stage-wise deficit irrigation was better than deficit irrigation throughout the growing season for improving yield productivity (Asres, 2023). Implementing a 25 percent deficit in water application has shown to enhance water productivity (WP), yielding 1.2 times the results achieved under normal irrigation practices. However, it is crucial to emphasize that the success of deficit irrigation relies on careful evaluation and irrigation scheduling to ensure an optimized and efficient use of irrigation water, as highlighted by Kirda in 2002.

Mulching Technology

To alleviate the water scarcity in agriculture, mulching has a vital impact as a water conserving technique in rain-fed cropping. It is very important primarily for preserving the soil moisture, preventing evaporation of soil and governing the soil temperature, which affects the food production. Mulching and micro- irrigation techniques can play a vital role in controlling the over and excessive irrigation. This practice assists to prevent weed growth, augments soil moisture, reduces soil erosion, maintains soil temperature, preserves soil structure, improves soil fertility and improvise soil biological regime (Prem *et al.*, 2020) There are several categories of mulching viz; organic mulching, inorganic mulching, natural mulching, synthetic mulching, surface mulching, upright mulching or vertical mulching, plastic mulching, rubber mulching, pebble or stone mulching, dust mulching, straw mulching, landscape fabric, live vegetative barriers etc. However, challenges arise with the disposal of plastic mulch, often neglected by farmers. Crops like groundnuts and tuber crops require careful separation of plastic during harvesting. While biodegradable plastic production is costly, educating farmers on proper usage and disposal is essential. Establishing a marketing system offering attractive prices for used plastics can incentivize degradable plastic mulch adoption. Additionally, promoting environmental sustainability suggests encouraging alternatives like burning agricultural residue or using organic mulching through natural farming practices. This approach ensures proper disposal and addresses environmental concerns associated with plastic mulch.

Conclusion

Investing in water security not only promises improvements in water availability and quality but also contributes significantly to the overall economic and social development of India. It is a collective responsibility that requires sustained efforts, cooperation and a shared commitment to building a resilient and sustainable water future for the nation. Enhancing water security in India demands a comprehensive strategy integrating policy, technology and behavioural solutions. A fundamental aspect involves bolstering water governance systems, establishing transparent policies, laws and institutions for the effective and equitable management and allocation of water resources. To curtail water wastage and enhance availability, it is imperative to adopt efficient water management practices, including rainwater harvesting, conservation and recharge.

For water security in agriculture and to maximize water productivity, the adoption of MISs alone will not suffice. We are at a juncture where we need to consider the entire spectrum of water resource issues. A collaborative approach involving the government, civil society, private sector and communities is crucial for successful water resource management. This approach can contribute not only to improved water availability and quality but also to the overall economic and social development of India.

References

[1] Bhagat, Arun Dattaray and Gholkar, Madhav D. and Powar, Ranjit V. (2023). Farmers' Perception of Mobile Application-Based Climate-Resilient Agromet Advisory Service Adoption and Effectiveness: Case Study in Maharashtra, India. https://ssrn.com/abstract=4471680 or http://dx.doi.org/10.2139/ssrn.4471680.

[2] Boursianis, A., Papadopoulou, M., Gotsis, A. Wan, S., Sarigiannidis, P., Nikolaidis, S. and Goudos, S. (2020). Smart Irrigation System for Precision Agriculture—The AREThOU5A IoT Platform. *IEEE Sensors Journal.* pp. 1–1. 10.1109/JSEN.2020.3033526.

[3] Castro-Pardo, M., Martínez, P. and Zabaleta, C. (2022). An initial assessment of water security in Europe using a DEA approach. *Sustainable Technology and Entrepreneurship.* 1, 100002. doi: 10.1016/j.stae.2022.100002.

[4] Das, B., Wani, S., Benbi, D., Muddu, S., Bhattacharyya, T. Mandal, B., Santra, P., Chakraborty, D., Bhattacharyya, R., Basak, N. and Reddy, N. (2022). Soil Health and its Relationship with Food Security and Human Health to Meet the Sustainable *Development Goals in India. Soil Security.* 8. 100071. 10.1016/j.soisec.2022.100071.

[5] Dina El Demerdash, Mohie El Din Omar, Mohamed Nour El-Din, Hesham El-Badry, Essam Aly and Doaa, A. El-Molla (2022). Development of a quality-based irrigation water security index, *Ain Shams Engineering Journal,* 13(5):101735.

[6] Eidi, A., Kazemieh, F. and Zarifian, S. (2020). Identifying the factors affecting the sustainable management of agricultural water resources from the perspective of farmers (case study: wheat farmers in rural areas of Maragheh). *J. Agric. Knowledge Sustain. Prod.* 13, 311–326. (Persian). Available online at: https://www.sid.ir/fa/journal/ViewPaper.aspx?id=571436.

[7] FAO (2017). Water for Sustainable Food and Agriculture. A Report Produced for the G20 Presidency of Germany. Rome: FAO. Available online at: http://www.fao.org/nr/water/docs/wateratagIance.

[8] Gruère, G. and Shigemitsu, M. (2021). Measuring Progress in Agricultural Water Management. Paris: OECD.
https://doi.org/10.1016/j.asej.2022.101735.

[9] Kirda, C. (2002). Deficit irrigation scheduling based on plant growth stages showing water stress tolerance. (In): FAO, Water Reports-22: Deficit irrigation practices (ISBN 92-5-104768-5), pp. 3–10.

[10] Ligalem, Agegn Asres (2023). Alternative Techniques of Irrigation Water Management for Improving Crop Water Productivity, *Reviews in Agricultural Science,* 11:36–53. https://doi.org/10.7831/ras.11.0_36, https://www.jstage.jst.go.jp/article/ras/11/0/11_36/_article/-char/en, Abstract:

[11] Malekian, A. (2017). Water Security Perception and Conservation Behaviors Among farmers In Fars Province. (Doctoral dissertation, University of Shiraze, Shiraze, Iran).

[12] Miras-Avalos, J.M., Perez-Sarmiento F., Alcobendas R., Alarcón J. J., Mounzer O. and Nicolás E (2016). Using midday stem water potential for scheduling deficit irrigation in mid-late maturing peach trees under Mediterranean conditions. *Irrigation Science,* 34(2): 161–173.

[13] Pandya, P.A. and Dwivedi, D.K. (2016), Constraints in Adoption of Drip Irrigation, *Advances in Life Sciences*, 5(6): 2405–2411.

[14] Pandya, P.A. and Gontia, N.K. (2023). Early crop yield prediction for agricultural drought monitoring using drought indices, remote sensing and machine learning techniques. Journal of Water and Climate Change. 14(12): 4729–4746. https://doi.org/10.2166/wcc.2023.386

[15] Pathak, A.R., Rank, H.D. and Pandya, P.A. (2021), Approaches for Improving Crop Water Productivity, Special Reference to Gujarat, Shodh Chintan, Innovative Approaches for Enhancing Water Productivity in Agriculture Including Horticulture, *ASM Foundation,* New Delhi, Vol. 12 & 13, May 2021, ISBN 978-81-932266-7-4.

[16] Prem, P., Ranjan, P., Seth, N. and Patle, G.T. (2020). Mulching Techniques to Conserve the Soil Water and Advance the Crop Production—A Review. Current World Environment. Special Issue (Sustainable Mining). DOI: http://dx.doi.org/10.12944 /CWE.15.Special-Issue1.02

[17] Rathore, V.S., Nathawat, N.S., Bhardwaj, S., Yadav, B.M., Santra, P., Kumar, M., Shekhawat, R.S., *et al.* (2022). Alternative cropping systems and optimized management practices for saving groundwater and enhancing economic and environmental sustainability. *Agricultural Water Management.* 272:107840. https://doi.org /10.1016/j.agwat.2022.107840.

[18] Saha, D., Sikka, A.K. and Goklani, R. (2022), Artificial recharge endeavours in India: A review, *Water Security,* 16, 100121, https://doi.org /10.1016/j. wasec.2022.100121.

[19] Scott, C.A., Zilio, M.I., Harmon, T., Zuniga-Teran, A., Díaz-Caravantes, R., Hoyos, N., *et al.* (2021). Do ecosystem insecurity and social vulnerability lead to failure of water security? *Environmental Development,* 38, 100606. doi: 10.1016 /j.envdev.2020.100606.

[20] Taylor, A. (2015). Large-Scale Land Acquisitions in Tanzania: A Critical Analysis of Their Implications on Water Security (Master dissertation), Dalhousie University Halifax, Nova Scotia.

[21] United Nations (2022). World Population Projected to Reach 9.8 Billion in 2050 and 11.2 Billion in 2100.

[22] Valizadeh, N., Esfandiyari Bayat, S., Bijani, M., Hayati, D., Viira, A.H., Tanaskovik, V., *et al.* (2021). Understanding farmers' intention towards the management and conservation of wetlands. *Land* 10, 860. doi: 10.3390/land10080860.

[23] World Bank (2019a). Annual Freshwater Withdrawals, Agriculture (% of Total Freshwater Withdrawal).

[24] World Bank (2019b). Level of Water Stress: Freshwater Withdrawals as a Proportion of Available Freshwater Resources.

Water Scarcity and Security in Agriculture: Challenges and Resolutions in India

Murari Lal Gaur[1]

ABSTRACT: Water is the most crucial input for agricultural production, playing a pivotal role in ensuring food security and sustaining livelihoods, particularly in countries like India where agriculture is a primary economic activity. However, water scarcity, inefficient water management practices, pollution and climate change pose significant challenges to water security in agriculture. This paper aims to explore the dynamics of water security in agriculture in India, examining the challenges faced and proposing sustainable solutions to mitigate these challenges. It also incorporates food for thought on critical issues like major obstacles in agricultural water management, climate change impacts on water security and scarcities, projected water securities for next few decades and also some of the strategic adaptations for such scarcities with relevant elaborations on gauged efforts, reasoning and solution matrix. Lastly, role of digital tools and technologies and related inventions & interventions towards smart water management in farming sectors are made part of writeup, to project some of the latest inventions, tools, technologies and potential options to overcome water securities and scarcities in agricultural and allied sectors. A summarized elaboration is offered at last with most updated references of key nature.

Keywords: Water Scarcity, Water Securities, Water Productivities, Management, Technologies.

Introduction

Water security in agriculture poses significant challenges in India due to various factors such as population pressure, climate variability, inefficient water management practices and socio-economic disparities. Water security in agriculture is indispensable for ensuring food security, alleviating poverty and fostering sustainable development in India.

---

[1] Senior Professor (SWCE), Anand Agricultural University Anand-388110 Gujarat, India,

Addressing the challenges of water scarcity, inefficient water use, groundwater depletion and pollution requires concerted efforts from the government, civil society and the private sector. By promoting sustainable water management practices, investing in infrastructure and fostering innovation, India can enhance water security in agriculture, thereby ensuring the resilience and sustainability of its agrarian economy for future generations.

There remain enormous contests in above regards which revolves around factors like water scarcity, its unequal distribution, groundwater (GW) depletion, climate change impacts, inefficient water management, water-based pollution, deficits in water governance & regulatory frameworks, limitations in regards to access for irrigation and many others. As a result of increasing in human water demand driven by rapid global population growth, economic development and rising living standards, water scarcity is further becoming a major threat to the sustainable development of human society, as it may result in multiple social and environmental issues, such as drinking water shortage, reduction of crop production and water quality degradation (Hanasaki *et al.,* 2018). Furthermore, global water scarcity is likely to change along with present and future climate change and socio-economic development. Therefore, it is essential to understand local & global water scarcity conditions as a prerequisite for developing water-based planning and management policies.

Current water scarcity projections are mostly based on water availability and water demand simulations from global hydrological models (GHMs), considering the impacts of both future climate change and socio-economic development (Hoekstra, 2014). Furthermore, a wealth of studies has always addressed the role of multiple adaptive options in coping with water scarcity conditions, e.g., irrigation efficiency improvement, inter-basin water transfer infrastructures (Yin *et al.*, 2022), seawater desalination and virtual water trade (Pastor *et al.*, 2022). Previous water scarcity assessments often compared water demand with water availability to assess water scarcity at the local, regional and global scales. Water demand was estimated as the sum of water demands from agricultural, domestic and industrial sectors in most studies (Shekh and Gaur, 2013), whereas water availability used to be calculated in different ways, such as from runoff, natural river discharge and the difference between natural river discharge and environmental flow. The allocation of available water resources between upstream and downstream regions is another important issue for water scarcity assessments. Different treatments of water abstraction in upstream areas may affect water scarcity assessment to a significant extent.

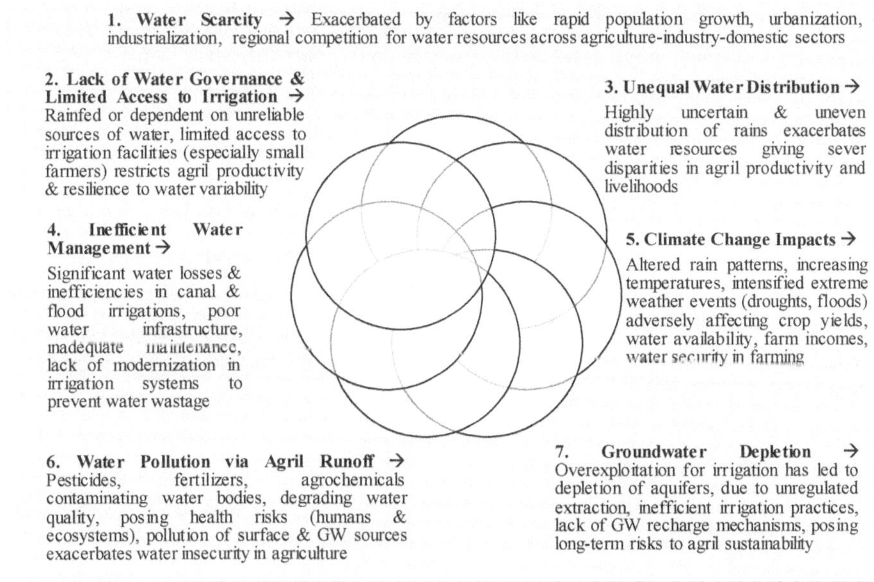

Figure 1: Unpacking the Challenges of Agricultural Water Scarcity

Despite significant investments in irrigation infrastructure, a large proportion of agricultural land in India remains rainfed or dependent on unreliable sources of water. Limited access to irrigation facilities, especially for smallholder farmers, restricts agricultural productivity and resilience to water variability (Mevada *et al.*, 2018). Figure 1 illustrates a detailed description of 7 key challenges with relevant cause-effect analysis. The contents and logical frameworks depicted herein clearly evolve the reasoning with which these challenges exacerbate the extent and intensities of water-based scarcities in the farming sector.

Agricultural Water Security in India: A Vital Concern

Agricultural water security matters most in the Indian context due to its pivotal role in ensuring food security, sustaining rural livelihoods, driving economic growth, promoting environmental sustainability and enhancing climate resilience. Addressing water scarcity challenges through innovative solutions, investment in water infrastructure and policy interventions is essential for safeguarding the future of agriculture in India and securing the well-being of millions of farmers and rural communities (Huang *et al.*,

2021). It is of paramount importance in India, where agriculture serves as the backbone of the economy, employing a significant portion of the population and contributing substantially to food security, rural livelihoods and economic growth. This elaborative explanation as provided in Figure 2, offers a comprehensive overview of why agricultural water security matters most in the Indian context, supported by updated reviews and logical indicators.

Major Obstacles in Agricultural Water Management

Key obstacles include inadequate infrastructure, inefficient water use practices, unsustainable groundwater extraction, fragmented governance structures and climate change impacts. Navigating the nexus of challenges within the agricultural water sector requires a holistic approach that addresses the interrelated nature of water scarcity issues, water stress indicators and water-based risks. By prioritizing sustainable water management practices, enhancing water governance mechanisms and fostering climate resilience, stakeholders can mitigate the impacts of water scarcity on agriculture and build a more secure and resilient food system for future generations (Liu, J. *et al.,* 2017). The obstacles, as narrated below hinder the effective management and allocation of water resources, exacerbating water scarcity issues and compromising agricultural productivity and livelihoods.

Water Scarcity Issues: Water scarcity arises when demand exceeds available water resources, leading to competition among various agrarian sectors, industry and domestic use.

Implications: In agricultural contexts, water scarcity manifests as reduced crop yields, loss of income for farmers, increased food insecurity and environmental degradation.

Water Stress Indicators: Water stress indicators assess the pressure exerted on freshwater resources by human activities, considering factors such as water availability, population growth and water demand. High levels of water stress indicate unsustainable water use practices, posing risks to agricultural production, ecosystem health and socio-economic stability.

Water-Based Risks: Water-based risks encompass a range of hazards, including floods, droughts, water pollution, salinization and depletion of freshwater resources. These risks pose significant threats to agricultural water security, livelihoods and food systems, exacerbating vulnerability to climate variability and environmental degradation.

Figure 2: Applied & Operational Concerns of Water Security and Scarcity Issues in Farming Sector

Interrelatedness and Linkages: Water Scarcity and Stress: Water scarcity exacerbates water stress by limiting water availability for agricultural activities, exacerbating competition for limited water resources among various users.

Water Stress and Risks: High water stress levels increase the likelihood of water-based risks (droughts/floods, amplifying their impacts on agricultural productivity and resilience.

Obstacles and Water Scarcity: Inadequate infrastructure, governance challenges and unsustainable practices contribute to water scarcity issues, exacerbating water stress and compounding agricultural challenges.

Climate Change and Water Management: Climate change intensifies water scarcity and stress through altered precipitation patterns, increased temperatures and more frequent extreme weather events, further straining agricultural water management efforts.

Impacts of Climate Change on Water Security & Scarcity

Climate change is intricately linked with water security and scarcity, presenting significant challenges for agricultural and allied sectors. Variability in precipitation patterns, increasing temperatures and extreme weather events exacerbate water stress, posing threats to agricultural productivity, food security and rural livelihoods. The multifaceted impacts of climate change on water availability, quality and utilization in the agricultural sector along with adaptation strategies to mitigate these challenges are discussed here as under.

Changes in Precipitation Patterns: Climate change alters precipitation patterns, leading to changes in the timing, intensity and distribution of rainfall. Some regions experience more frequent and intense rainfall events, while others face prolonged dry spells and droughts. It has enormous impacts on agriculture. For example, the erratic rainfall disrupts agricultural planning and operations, affecting crop growth, irrigation scheduling and water management practices. Droughts reduce water availability for irrigation, leading to crop failures, loss of income and rural distress.

Rising Temperatures: Global warming increases temperatures, exacerbating evaporation rates and water loss from soil and surface water bodies. Higher temperatures also influence crop water requirements, affecting agricultural productivity. It too has big impacts on farming, as the elevated temperatures accelerate soil moisture depletion, increase crop water demand and aggravate water stress in agricultural landscapes. Also, the heat stress affects crop yields, livestock health and overall farm productivity.

Extreme Weather Events: Climate change augments the frequency and intensity of extreme weather events, including hurricanes, cyclones, floods and heatwaves. These events disrupt water availability, infrastructure and agricultural systems. As far as its real impacts on farming are concerned it comprises variety of situations and risks; like the floods inundate farmlands, destroy crops and degrade soil fertility, while heatwaves and cyclones cause heat stress, crop damage and infrastructure destruction. Such events disrupt supply chains, exacerbating food insecurity and rural poverty.

Water Quality Degradation: Climate change influences water quality through changes in temperature, precipitation patterns and hydrological cycles. Increased runoff, erosion and pollution from agricultural activities further degrade water quality in surface and groundwater sources. Looking towards its impacts, it clearly projects that the any of the poor water quality affects crop health, livestock production and human health, leading to reduced yields, contamination of food products and increased disease risks. Waterborne pathogens and pollutants pose significant challenges for agricultural sustainability.

The above challenges inevitably need to evolve demand driven solutions, for which optimum adaptation strategies remains inevitable. Some of the well proved strategies will remains as illustrated in Figure 3. It involves multiple options that remain focused on better water managerial aspects, better crop selections and management, better water-based infrastructure and effectual water policy-based interventions.

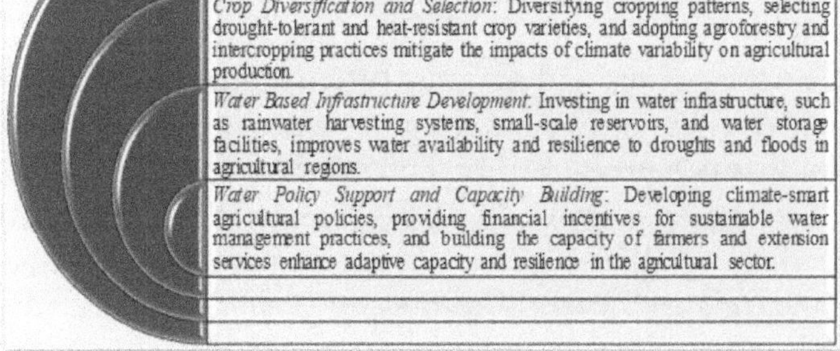

Figure 3: Demand-Driven Approaches to Mitigate Farming Water Scarcity

Projected Water Scarcity by 2050

As we navigate the complexities of projected water scarcity, it is imperative to prioritize water sustainability, resilience and equity to secure a water-secure future for generations to come. As we approach the year 2050, projections indicate that water scarcity will become increasingly prevalent and severe, posing significant challenges to agricultural sustainability, human well-being and ecosystem health. Rapid population growth, urbanization, climate change impacts and unsustainable water management practices contribute to the exacerbation of water scarcity worldwide. By 2050, the global population is expected to surpass 9 billion, further straining water resources as demand for food, water and energy escalates. Urbanization trends continue to accelerate, leading to increased water consumption, pollution and competition for limited water supplies.

Climate change worsens water scarcity through altered precipitation patterns, rising temperatures and more frequent extreme weather events. Shifts in hydrological cycles, prolonged droughts and intensified floods disrupt water availability and exacerbate water stress in vulnerable regions. Inefficient water-use practices, groundwater depletion and pollution from agricultural runoff, industrial discharge and urban wastewater degrade water quality and compromise ecosystem resilience. Inequitable water distribution and governance issues exacerbate water scarcity and exacerbate socio-economic disparities.

Water Footprints and Virtual Water Quantifications

Water footprints quantify the total volume of freshwater used directly and indirectly in the production, processing and consumption of goods and services. Agriculture accounts for the largest share of global water footprints, with water-intensive crops such as rice, wheat and cotton contributing significantly to water consumption (Quinteiro *et al.,* 2018). Virtual water quantifications assess the amount of water embedded in traded goods and commodities, reflecting the water footprint of products exchanged across regions and nations. International trade in virtual water allows water-stressed regions to import water-intensive goods, thereby relieving local water pressures but potentially transferring water scarcity burdens to exporting regions. By 2050, projected water scarcity poses a formidable challenge that requires urgent and concerted action at local, national and global levels. Addressing water scarcity demands a multifaceted approach that integrates sustainable water management practices, climate resilience strategies and equitable water governance frameworks. Investing in water-efficient technologies, promoting precision agriculture and adopting conservation practices can enhance water productivity and mitigate water scarcity risks in agriculture. Building resilience to climate change impacts through ecosystem restoration, drought-resistant crop varieties and adaptive water management strategies is essential for safeguarding water resources and agricultural livelihoods. Implementing integrated water resource management approaches, enhancing stakeholder participation and fostering international cooperation are crucial for ensuring equitable water allocation and addressing water scarcity challenges in a changing world.

Strategic Adaptation of futuristic Water Shortages

As the world copes with escalating water scarcity, future generations must adopt proactive strategies to live sustainably in water-limited environments. This entails understanding the gauging effects of water scarcity on agriculture and allied sectors, identifying the root causes of water crises and implementing a comprehensive solution matrix to ensure water security for agricultural sustainability and human well-being. Adapting to water shortages requires a multi-faceted approach that integrates technological innovations, policy reforms and behavioural changes to ensure sustainable water management practices in agriculture and allied sectors. By addressing above root causes of water crises and implementing a comprehensive solution matrix, future generations can live in harmony with water-limited environments, safeguarding agricultural sustainability, human well-being and ecosystem resilience for generations to come.

Gauging Effects of Water Scarcity on Agriculture

Reduced Crop Yields: Water scarcity leads to diminished irrigation supplies, impacting crop growth and yield potential.

Economic Losses: Farmers incur financial losses due to decreased productivity, higher input costs and reduced market competitiveness.

Food Insecurity: Water shortages disrupt food production, leading to food shortages, price volatility and nutritional deficiencies.

Rural Displacement: Water-stressed regions experience rural-to-urban migration as farmers seek alternative livelihoods, exacerbating urbanization pressures.

Major Reasons for Water Crisis

Population Growth: Rapid population growth strains water resources, increasing demand for agricultural, domestic and industrial purposes.

Climate Change: Altered precipitation patterns, rising temperatures and extreme weather events exacerbate water scarcity, affecting water availability and distribution.

Poor Water Governance: Inadequate policies, inefficient water management practices and fragmented governance structures contribute to water misallocation and wastage.

Unsustainable Agriculture: Water-intensive farming practices, such as flood irrigation and monoculture, deplete water resources and degrade ecosystems.

Pollution: Contamination of water sources by industrial effluents, agricultural runoff and urban waste reduces water quality and exacerbates water scarcity issues.

Solution Matrix

Efficient Water Use: Promote water-saving irrigation technologies like drip and sprinkler irrigation, adopt precision agriculture practices and optimize water use efficiency in crop production.

Sustainable Agriculture: Implement agroecological approaches, such as organic farming, conservation agriculture and crop rotation, to reduce water consumption and enhance soil health.

Integrated Water Management: Adopt integrated water resource management (IWRM) approaches, incorporating rainwater harvesting, groundwater recharge and watershed management strategies (Nyakundi *et al.*, 2022).

Policy Reforms: Enact and enforce water governance policies that prioritize equity, efficiency and sustainability, including water pricing mechanisms, water rights allocation and regulatory frameworks for water use.

Climate Resilience: Invest in climate-smart agriculture practices, develop drought-tolerant crop varieties and enhance adaptive capacity to mitigate the impacts of climate change on water availability.

Public Awareness and Education: Raise awareness about water conservation, promote community-based water management initiatives and provide training and capacity-building programs for farmers and stakeholders.

Water in Digital Era: Bridging Tradition & Technology

In the digital era, agricultural water management faces a dual challenge; leveraging cutting-edge technologies while integrating traditional knowledge systems to ensure sustainable and efficient water use. This elaborative illustration explores newer interventions that reimagine agricultural water management, blending smart wisdom with digital innovations to address contemporary water challenges while honouring age-old practices.

Digital Innovations in Water Management

Smart Irrigation Systems: Advanced sensors, IoT (Internet of Things) and AI (Artificial Intelligence) algorithms optimize irrigation scheduling, reducing water waste and enhancing crop yields.

Remote Sensing Technologies: Satellite imagery and drones provide real-time data on soil moisture levels, crop health and water stress, enabling precise water management decisions.

Mobile Applications: Farmer-centric apps offer personalized advice on irrigation, weather forecasts, crop selection and water-saving practices, empowering farmers to make informed decisions.

Integration of Traditional Knowledge

Indigenous Water Management Practices: Drawing from traditional wisdom, techniques such as rainwater harvesting, contour bunding and crop diversification promote water conservation, soil health and resilience to climate variability.

Community-led Water Governance: Participatory approaches based on local knowledge systems foster collective action, equitable water allocation and sustainable management of water resources.

Agroecological Farming Systems: Traditional agroecological practices, such as agroforestry, intercropping and integrated pest management, enhance water efficiency, biodiversity and ecosystem services.

Reinventing Traditional Wisdom with Digital Tools

Digital Knowledge Platforms: Online repositories of indigenous knowledge, best practices and success stories facilitate knowledge sharing, capacity building and cross-cultural exchange among farming communities.

Citizen Science Initiatives: Crowdsourcing platforms engage farmers in data collection, monitoring and research, enabling co-creation of knowledge and innovation in agricultural water management.

Blockchain Technology: Transparent and decentralized ledger systems enhance traceability and accountability in water transactions, promoting fair water governance and reducing water conflicts.

Salient Points of Concern

Access and Equity: Disparities in digital literacy, internet connectivity and access to technology may exacerbate inequalities, marginalizing small farmers and rural groups.

Data Privacy and Security: Safeguarding farmer data, ensuring data ownership rights and addressing concerns regarding data privacy and security are critical in digital age.

Cultural Preservation: Balancing technological advancements with the preservation of indigenous knowledge systems and cultural heritage is essential for respecting traditional wisdom and fostering cultural resilience.

Latest Innovations in Addressing the Water Crisis

While innovative technologies such as the Super Imploder, Atmospheric Water Generators and Nano-Irrigation offer promising solutions to address water scarcity and enhance water productivity in farming, their widespread adoption and long-term implications require careful evaluation, validation and consideration of socio-economic, environmental and ethical considerations. Collaboration between researchers, policymakers, farmers and technology developers is essential to ensure that these inventions contribute positively to sustainable agriculture, water conservation food security for present and future generations.

Super Imploder Technology: The Super Imploder is a water treatment device that claims to enhance water quality, increase dissolved oxygen levels and reduce surface tension through implosion technology. As far as its working principle is concerned, the device utilizes a vortex-based implosion system to enhance water structuring and energy transfer, purportedly improving water absorption and nutrient uptake in plants.

Probable Implications for Farms: It includes enhanced crop growth, as treated water promotes healthier plant growth, increases crop yields and reduces the need for fertilizers and agrochemicals. Also, the improved water quality may enhance soil microbial activity, soil structure and nutrient availability, contributing to soil fertility and long-term sustainability. Water savings may also be easily attained via increased water absorption by plants, which in turn may lead to reduced irrigation requirements and water usage, potentially benefiting water-stressed farming regions. While some anecdotal evidence suggests positive impacts on plant growth and soil health, independent scientific studies validating the efficacy and long-term effects of Super Imploder technology are limited.

Atmospheric Water Generators (AWGs): They extract moisture from the atmosphere and condense it into potable water, offering a decentralized and renewable water source for agricultural and domestic use. As far as their working principle is concerned, it utilizes refrigeration or desiccant technologies to cool air, causing water vapour to condense into liquid water, which is then filtered and stored for consumption. It has multiple probable implications for farms, out of which few could be narrated as below,

Water Security: AWGs provide an alternative water source for irrigation, livestock watering and farm operations, reducing dependence on traditional water supplies such as GW and surface water.

Climate Resilience: AWGs are particularly beneficial in arid and semi-arid regions where water scarcity is prevalent, offering a sustainable solution for mitigating climate-related water risks.

Energy Requirements: The energy consumption of AWGs may vary depending on the technology used, with refrigeration-based systems requiring electricity (potentially added operational costs).

Considerations: Factors such as humidity levels, air quality and initial investment costs should be evaluated when considering the adoption of AWGs on farms.

Nano-Irrigation Technologies: It deliver water and nutrients directly to plant roots at the nanoscale, optimizing water use efficiency and minimizing water losses through evaporation and runoff. It utilizes nanostructured materials, such as nanomembranes and nano-sensors, to regulate water delivery, monitor soil moisture and enhance nutrient uptake by plants. It has enormous probable implications for farms like;

Precision Irrigation: Nano-irrigation enables precise control over water application, matching water supply to plant demand and minimizing wastage, leading to improved crop yields and resource efficiency.

Reduced Environmental Impact: By minimizing water runoff and leaching of fertilizers and pesticides, nano-irrigation technologies contribute to environmental sustainability and reduce agricultural pollution.

Technological Challenges: The development and implementation of nano-irrigation systems require specialized expertise, infrastructure and investment, posing challenges for widespread adoption, especially in smallholder farming contexts.

Regulatory frameworks, safety concerns related to nanomaterials and long-term environmental impacts warrant further research and assessment before widespread deployment of nano-irrigation technologies in agriculture.

Summary and Conclusion

Water security and scarcity in the agricultural sector pose significant challenges to food production, rural livelihoods and environmental sustainability. Factors such as climate change, population growth, unsustainable water management practices and inadequate infrastructure contribute to water stress in farming communities worldwide. Addressing these challenges requires a multi-faceted approach that integrates technological innovations, policy reforms and community engagement to optimize water use efficiency, enhance resilience to climate variability and promote equitable water allocation. Innovative solutions such as precision irrigation technologies, rainwater harvesting systems and sustainable land management practices offer opportunities to improve water productivity, conserve water resources and ensure the long-term viability of agriculture. Issues becomes much complex that demand urgent attention and concerted action from governments, stakeholders and communities. By prioritizing sustainable water management practices, investing in water infrastructure and fostering innovation and collaboration, we can address the root causes of water scarcity, mitigate its impacts on agricultural productivity and build resilience to future water challenges. As we strive to achieve a water-secure future for all, it is essential to recognize the intrinsic linkages between water, food, energy and environmental sustainability and work towards integrated solutions that benefit both people and the planet. Through collective efforts and shared responsibility, we can ensure that agriculture remains a source of livelihood, food security and environmental stewardship for generations to come.

Among indicators, the floods and droughts pose biggest challenges with significant socio-economic and environmental implications at both regional and global scales. Understanding the vulnerability of different regions to these hazards is essential for developing targeted adaptation and mitigation strategies. By addressing underlying drivers, enhancing preparedness and fostering collaboration, countries can build resilience and reduce the impacts of floods and droughts on communities and ecosystems. Addressing above challenges requires comprehensive strategies that integrate sustainable water management practices, climate-resilient agriculture, equitable water governance and community participation. Investing in water-saving technologies, promoting efficient irrigation methods, enhancing water infrastructure and empowering farmers with knowledge and resources are essential for ensuring water security in agriculture in India. Moreover, fostering interdisciplinary research, strengthening policy frameworks and promoting cross-sectoral collaboration are crucial for addressing the multifaceted challenges of water security in agriculture and ensuring sustainable food production and livelihoods in the country.

Further climate change poses formidable challenges to water security and scarcity in agricultural sectors, with far-reaching implications for food production, rural livelihoods and environmental sustainability. By understanding the complex interactions between climate change and water resources and implementing adaptation strategies that enhance resilience and sustainability, stakeholders can mitigate the impacts of climate variability and ensure the long-term viability of agricultural systems in a changing climate. In the digital era, the convergence of tradition and technology holds immense potential for transforming agricultural water management. By embracing newer interventions that integrate smart wisdom with digital innovations, we can reimagine water stewardship, promote sustainable agriculture and build resilience to water challenges in a rapidly changing world. However, it is imperative to address points of concern and ensure that technological advancements are inclusive, ethical and culturally sensitive, honouring the wisdom of the past while embracing the opportunities of the future. A location-specific water accounting system is indeed the need of the hour for wiser conservation and management of agricultural water. By providing stakeholders with accurate, timely and context-specific information on water availability, allocation and use, such a system empowers decision-makers to implement targeted interventions that optimize water resources, enhance water security and promote sustainable agriculture for the benefit of present and future generations.

References

[1] Hanasaki, N., Yoshikawa, S., Pokhrel, Y. and Kanae, S. (2018). A quantitative investigation of the thresholds for two conventional water scarcity indicators using a state-of-the-art global hydrological model with human activities. *Water Resources Research,* 54, 8279–8294.

[2] Hoekstra, A. Y. (2014). Water scarcity challenges to business. Nature Climate Change, 4(5), 318–320.

[3] Huang, Z., Liu, X., Sun, S., Tang, Y., Yuan, X. and Tang, Q. (2021). Global assessment of future sectoral water scarcity under adaptive inner-basin water allocation measures. *The Science of the Total Environment*, 783, 146973.

[4] Liu, J. *et al.,* 2017. Water scarcity assessments in the past, present and future, *Earth's Future,* 5(6), 545–559.

[5] Mevada, K.D., Gaur, M.L. and Shah, Tushar (2018). Water management Chapter 6 In: Book entitled Strategies for doubling the farmers income in India (Eds: Sanjay Prashad, N.C. Patel and Mbuya, Odemari S.) A Strategic R&D Book Pub: Anand Agril. Univ Anand Gujarat India.

[6] Nyakundi, R., Nyadawa, M. and Mwangi, J. 2022. Effect of recharge and abstraction on groundwater levels, Civil Eng. J. 8 (5), 910–925.

[7] Pastor A.V., Biemans H., Franssen W., Gerten D., Hoff H., Ludwig F. and Kabat P. (2022). Understanding the transgression of global and regional freshwater planetary boundaries Phil. *Trans. R. Soc.* A.38020210294202 10294

[9] Quinteiro, P., Ridoutt, B.G., Arroja, L. and Dias, A.C. (2018). Identification of methodological challenges remaining in the assessment of a water scarcity footprint: a review, *Int. J. Life Cycle Assess,* 23(1), 164–180.

[10] Shekh, A.M. and Gaur, M.L. (2013). Differential Impacts of Climate Change on Water Resources: An overview of Gujarat. Chapter No. 1 (pp. 1–19) in Book entitled 'Water Management Research in Gujarat' 171 Pages Pub: SMRU, Navsari Agril University, Navsari Gujarat, ISBN: 978-81-923919–1-8.

[11] Yin Dongqin, Qian Li, Xiang Li and Ruirui, Zhu (2022). Annual water deficit in response to climate variabilities across the globe, *Environmental Research Letters,* Volume 17, Number 5. DOI 10.1088/1748-9326/ac5d9b

Water Resilient Agriculture

Suresh Acharya[1]

ABSTRACT: Water resilient agriculture (WRA) redefines farming with a circular water approach (reduce, recycle, reuse). Smart irrigation and real-time data empower data-driven water management. Eco-friendly practices ensure long-term sustainability. For resilience, WRA integrates crop diversification (e.g., water-efficient CAM-transited groundnuts, flood-tolerant rice). Research explores transformative innovations like "Carbon-Sucking Plants" that capture carbon and boost water efficiency. Automation offers wider adoption of zero-tillage practices, conserving soil and water. Emerging technologies (nanotech, VRT) and AI-powered WRA systems further enhance water management precision. These advancements, coupled with responsible practices, pave the way for global water-resilient agriculture.

Introduction

Food and water are inseparable and the challenge is to produce more with less water. India, with 18% of the global population, only possesses 4% of the world's freshwater resources. Usable water stands at 28%, 1.123 lakh crore cubic meters (LCCM) of the total 4 LCCM, entailing 0.69 and 0.433 LCCM from surface and underground sources, respectively. The formidable demand of 1.147 LCCM make it still dicer and highlights the need for a delicate balance between supply and consumption. The annual per capita water availability has dropped dramatically over the last seven decades from 5177 L to 1341L, categorizing India as water stressed, which is further exacerbated by 51% of the area being rainfed. Agriculture exacerbates the problem by consuming 84% of water with a water use efficiency (WUE) of 0.38, with significant water losses ranging from 20–40% in intensive agricultural regions. The situation is further complicated by a substantial nexus of evapotranspiration losses (1500–3500 mm) coupled with least growth period (LGP) of less than 90 days in most parts of the country, leading to severe stress during the reproductive phase. The water resilient agriculture encompasses practices like precision and drip irrigation, rainwater

---

[1] AGROCEL Industries Private Limited.

harvesting and selecting drought-tolerant crops. Conservation practices like cover cropping and tillage enhance soil health, improve water storage and reduce environmental impacts. India has excelled across different dimensions of water management.

Plants as Automated Super Machine

For water stewardship in agriculture, the 'SW-MEALS' (Soil-Water-Microbes-Energy-Agriculture-Livelihood Security) nexus advocates a holistic approach. Plants, intricate systems of resilience, function like "automated sensor-fitted super machines," adapting dynamically to environmental changes. In response to water scarcity, plants adjust by closing stomata, shedding leaves and growing deeper roots. Despite their resilience, the delicate balance between water availability and plant vitality is manifested in responses like wilting during extreme heat. Beyond their role as 'Super Machines,' plants act as nature's engineers, converting carbon dioxide into carbohydrates using water and sunlight, crucial for ecological balance.

At the heart of plant functionality is the enzyme Rubisco, which has dual affinity for carbon dioxide (CO_2) and oxygen (O_2). While CO_2 interaction boosts productivity by fixing 3 mole-carbon, challenges arise with exhaustive interactions with O_2 when CO_2 concentration becomes low, impacting plant efficiency. Nature introduced two adaptive models to prevent Rubisco-O_2 interaction: physical separation (C4 plants for warm conditions) and temporal separation (CAM plants for dry and warm conditions). Notably, CAM with 1/5 and 1/4 of water usage compared to C3 and C4 highlights its water-saving advantages. Despite, 89% are C3 plants suggests that our ancestors prioritized sustenance over water-saving. Obviously, the water stewardship underscores understanding the inter-connectedness of natural adaptation under different ecosystems for preserving the intricate web of life.

Super Machine Treated as Water Pump

Human-centric perspectives have inadvertently disrupted the delicate balance of the 'Super Machine' in various ways, transforming water-efficient plants into water-intensive crops. Examples include converting water-efficient xerophytic deep-rooted plants like cotton and castor into water-intensive crops. Regions like Punjab, extensively cultivate rice, while arid Kachchh is taking to water-intensive banana and papaya cultivation.

In semi-arid North Gujarat, substantial water resources are required for fodder cultivation for the classic dairy sector developed in the region. Even technological advancements, as observed in Deesa with sprinkler irrigation, face challenges, leading to water overflow due to a lack of ownership, irresponsive use and subsidized electricity.

The *'Amrit kaal'* emphasizes addressing these challenges by recognizing water as a shared resource across technological, social, cooperative and economic levels. Embracing best practices, from tillage to gene editing, is crucial for enhancing the efficiency of the 'Super Machine,' especially considering persistently low WUE in India, despite enormous government initiatives like 'More Crop Per Drop' and collateral programmes. Despite super machine has done exceedingly well achieving unprecedented milestones of 33.05 crore tons of food grains and 35.19 crore tons of fruits and vegetables production during 2022–23. Pulses production has surged to 2.89 crore tons from a previous range of 1.3–1.4 crore tons, while oilseeds production, a historical concern, has reached 4.15 crore tons. This highlights a misconception, treating plants as water pumps instead of super machines. Despite our shortcomings, plants have not failed us, though the ecological balance is at risk. Embracing innovation and scientific tools is imperative to address these challenges and accomplish *'Water Security in Agriculture'*.

Water Budgeting

Water budgeting at the crop and plant level requires precision in assessing water needs for each crop under varying soil and growth conditions. Advanced technologies, including AI, nanotechnology, tensiometers and the calculation of the IW/CPE ratio, etc, integrate various weather conditions into the process. Implementing participatory water management engages farmers and communities in decision-making regarding water allocation and infrastructure development, making them responsible for preparing *'Demand-side Water Budgets'* aligned with available *'Supply-Side Resources'* and different water losses. Water budget planning should be rooted in the Least Growth Period (LGP). Technology integration, including sensors, drones, AI and data analytics, enables precise water budgeting from local water bodies to river basin management, focusing on water use efficiency, irrigation optimization and weather pattern prediction. Strategies such as crop diversification, intercropping, early warning systems, fair water pricing mechanisms and knowledge sharing further contribute to effective water budgeting. Gujarat is a pioneer in water management, showcasing innovative models from village-level initiatives like Dhandi village in Anand district, where an efficient water budgeting model is employed to draw water from the aquifer based on agricultural needs. The Rukmavati River basin management in Kachchh highlights the transformative impact of water harvesting on agriculture, yielding positive outcomes in a region facing challenges of water scarcity and degraded water quality, including salinity.

3Rs Crucial for WRA

Water-resilient agriculture hovers around the 3Rs—reduction, recycling and reusing water—prompting a shift from traditional consumption models to circular paradigms. Repurposing wastewater as an irrigation source eases strain on freshwater resources, fostering a symbiotic relationship with nature and contributing to ecosystem restoration. Innovative technologies for the 3Rs, including smart irrigation systems, advanced water treatment and upgraded reservoir monitoring, enhance precision in water use, prevent losses and strengthen resilience against adverse weather. Technologies such as satellite imaging, artificial intelligence, IoT and drones provide real-time information on water levels and soil conditions, contributing to efficient water management in farming. Ensuring eco-friendliness involves sediment management, biodiversity preservation and reducing the carbon footprint of structures. Desalination, from traditional methods to modern techniques like reverse osmosis, holds promise for securing water, though addressing environmental concerns, such as brine disposal and energy consumption, is crucial.

Conceiving Beyond 3Rs

Water-resilient agriculture, guided by the principles of the 3Rs—reduction, recycling and reuse of water—customizes water usage to specific crop needs. India's emphasis on crops like *'Shrianna'* or millets reflects a commitment to addressing global warming. Persistent endeavors to tackle threats from water scarcity necessitate ongoing re-assessment beyond the 3Rs. The ground breaking development of Crassulacean Acid Metabolism (CAM) varieties in groundnut showcases a significant water-use reduction compared to traditional C3 and C4 photosynthesis systems. Three CAM varieties were tested for salinity at Mandvi, with DGRMB5 and DGRMB19 officially registered for salinity tolerance by NBPGR, New Delhi. The emergence of CAM plants under stress suggests potentials of selecting CAM types for enhancing water efficiency in different crops.

The shift towards submergence followed by extended dry spells has spurred the development of dual-responsive varieties that can withstand both drought and submergence. Research initiatives are underway, focusing on selecting submergence-resistant varieties even for dryland crops like chickpea and pigeonpea. The *'Ranidhan'* variant, featuring the SUB1A QTL from traditional varieties and three BB resistance genes (Xa21, xa13 and xa5), demonstrates resilience, enduring submergence for a fortnight and holds promise in rainfed lowlands, especially in eastern India. Introducing genes from *Cyanobacteria* and red algae into tobacco plants has resulted in 27% larger growth and improved resilience to variable water conditions, including dry, normal and submergence scenarios. Transformative innovations include turning plants into 'Carbon-Sucking Super Machines,' storing absorbed CO_2 in roots for carbon storage and extracting water from deeper layers, enhanced with useful microbes and 50 identified genes. Investigating the resilience of the traditional wheat variety Lok 1 revealed that its adaptation to an extensive sowing time window from October to January is attributed to its unique feature of double root length compared to normal varieties.

Zero Tillage

Misplaced carbon (more in environment and less in soil) is the root cause of climate change and poor Resource Use Efficiency (RUE). Soil carbon correlates directly with organic matter, water holding capacity and permanent wilting point, influencing available water. Zero tillage demonstrates advantages in soil carbon content, increasing by 235% and sequestering 3.34 times the organic carbon compared to initial levels. In contrast, deep and conventional tillage only showed increases of 1.53 and 1.74 times, respectively. Additionally, zero tillage enhances nitrogen fixation, nodules per plant, earthworm activity, soil quality and Water Use Efficiency (WUE) without soil erosion, aligning with Water-Resilient Agriculture (WRA) principles.

Despite its considerable benefits, farmers often view Zero-Tillage as impractical. The automation of zero tillage presents a promising frontier for agricultural innovation. The trio of robots—*Tom, Dick and Harry*—seamlessly connected to a computer equipped with WIMA "AI," introduces intelligence to each plant. Tom conducts field mapping, Dick utilizes lasers to zap weeds and Harry performs seed drilling without the need for traditional tillage. Technology integration of zero tillage through KYS (Know Your Soil) involves three software components; (i) BeCRop, which employs DNA sequencing and AI tools to understand the soil microbiome; (ii) Gheom, providing predictive insights into the effects on the soil microbiome to aid informed decisions; and (iii) 23&MeForSoil, anticipating how microbiome interventions may influence soil productivity, quality and sustainability.

Responsible Water Management

As a vital component of responsible water management in agriculture, AGROCEL Industries Private Limited (AIPL), as part of its Corporate Social Responsibility (CSR) initiatives, deepened 12 ponds in Kachchh and 8 ponds in Bhavnagar during 2023. The total capacities of these ponds were 65,483 CuM and 60,033 CuM, respectively. In Kachchh and Bhavnagar, 88 and 51 Individual Rainwater Harvesting Structures (RWHS) were implemented, harvesting 706.44 CuM and 468.80 CuM of water, respectively. Additionally, 8 Institutional RWHS projects were executed, harvesting 263.00 CuM of water. As part of the Kaka (Sh Kantisen C. Shroff) Centenary Programme on Jal Mandir, 216 Jal Mandir structures were successfully established, with an estimated total water harvesting of 8,76,660 CuM.

Anti-evaporation techniques, including floating covers, chemical additives and windbreaks, effectively minimize water loss. Efficient water utilization techniques, such as drip irrigation, sprinkler irrigation, subsurface irrigation and laser irrigation, optimize water use by addressing challenges like percolation, runoff and evaporation. These methods enhance plant growth and yield. Drainage and leaching are essential for sustainable water management, especially amid climate challenges. Effective drainage prevents waterlogging, while leaching is crucial for controlling soil salinity, adapting to changing precipitation patterns.

Mulching, involving the use of organic or synthetic materials, conserves soil moisture, regulates temperature, suppresses weeds and improves soil structure. This practice significantly contributes to increased crop yields, reduced irrigation needs and enhanced farm income. Cover cropping, implemented during main crop dormancy, protects soil from erosion, enhances structure, suppresses weeds and acts as a natural fertilizer. Improved soil health leads to increased yields, reduced dependence on external inputs and higher farm income.

Agroforestry, integrating trees or shrubs into agriculture, diversifies farming systems, providing multiple income streams. Trees contribute to timber, fruit and fodder, enhancing biodiversity and promoting sustainable land management. Agroforestry fosters sustainable agriculture and supports farmers' economic well-being by optimizing resource use.

Plant-Microbes Interaction

NCCSD conducted experiments in three districts, focusing on the effectiveness of *Pseudomonads* and *Bacillus* in cotton. Vermiwash cocktails of microbes and PGR also exhibited positive results. Our experiment involving *mycorrhizal* fungi combined with plant peptides and humic acid has proven to be effective. In a notable innovation in Hyderabad, soil microbes were found to produce a weak electric current in the soil. Electrodes were developed to generate a low voltage current, simulating the one produced by microbes. Initial experiments demonstrated increased leaf area, plant height and improved partition of vegetative and reproductive phases. Recognizing the sensitivity of microbes to shelf life, storage and logistics, biocapsules have been introduced. Gelatine encapsulates microbial strains, including mycorrhizal fungi that enhance phosphorus uptake by up to 400%, promoting root growth and water absorption. Encapsulated humic acid increases soil water holding capacity by up to 20%, reducing irrigation requirements and enhancing drought tolerance. These examples highlight the multifaceted benefits of biocapsules in promoting water-resilient agriculture.

New Technologies

Emerging technologies are revolutionizing water-resilient agriculture by offering precise and efficient solutions. AGROCEL's experiment with Evaloc, a fatty acid molecule, in Ludiya Village, Bhuj, Kachchh, Gujarat, showed a 35% reduction in water evaporation when applied alone and 43% when combined with polythene. Antitranspirants, like Kaolin clay and Polyethylene glycol, reduce transpiration by up to 50%, aiding crop yields in drought conditions. Signalling molecules, such as Abscisic acid and Chitosan, play a vital role in plant adaptation to water stress. Paclobutrazol enhances stress resistance and water use efficiency. Silicon, in the form of silica, improves stress tolerance and drought resistance, forming a physical barrier to reduce water loss through transpiration and increase overall water use efficiency. These tools demonstrate a multifaceted approach to enhance water resilience in agriculture.

Nanotechnology introduces nanoscale biosensors to detect plant stress and pollutants, while nanomaterials remediate contaminated water.

Additionally, plant nanobionics enhances Water Use Efficiency (WUE). As nanotechnology evolves, its potential applications in agriculture are boundless. Variable Rate Technology (VRT) offers a targeted approach using GPS, sensors and data analytics. By generating 'Prescription Maps,' VRT directs irrigation equipment for tailored water application based on specific field or plant requirements. VRT is bifurcated into Map-Based and Sensor-Based categories, utilizing varied data sources from satellite imagery to real-time information for precise irrigation mapping. Complementing these innovations, Hyperspectral Technology captures nuanced changes in crops, soil and water reflectance. With capabilities like push broom, whisk broom and sequential scanners, this technology, often mounted on drones or satellites, facilitates meticulous monitoring of water parameters, ensuring quality and early detection of deficiencies. Together, these advancements enhance decision-making, fostering water-resilient agricultural practices.

AI-powered Water-Resilient Agriculture (WRA) integrates GPS, AI algorithms and advanced data analytics to provide real-time insights into soil moisture levels, crop conditions and weather patterns. This allows for informed decisions that optimize Water Use Efficiency (WUE). AI-driven systems meticulously analyse vast amounts of data to optimize irrigation schedules, ensuring precise water application based on actual needs. This minimizes water wastage and promotes sustainable water use. Predictive analytics, a cornerstone of AI-powered agriculture, enables farmers to anticipate and proactively mitigate the impact of climate changes, fostering overall agricultural resilience. Silicon Valley's 'Climate AI' provides customized data on climate, water and soil variability for the upcoming season, aiding pre-emptive decision-making. Additionally, drone technology has transformed agricultural practices by offering a valuable aerial perspective for efficient crop monitoring, swiftly identifying issues and enhancing decision-making processes.

Conclusion

Addressing the complexities of the SW-MEALS (Soil-Water-Microbes-Ecology-Agriculture-Livelihood Security) nexus in water-resilient agriculture requires a comprehensive strategy. Balancing the 'Demand Side' with the 'Supply Side,' weather conditions and the natural resource base is crucial. The synergy between traditional methods and cutting-edge technologies is indispensable for a sustainable and resilient agricultural future. A participatory approach involving farmers and communities in local water budgeting, with incentives for water savings, is vital for promoting responsible water use. Innovative technologies, including smart irrigation systems, antitranspirants, anti-evaporation tools and biocapsules, play a significant role in optimizing water management. Achieving water-resilient agriculture amid climate change requires a harmonious blend of traditional wisdom and innovative technologies, emphasizing the need for comprehensive strategies to enhance water resilience.

Enhancing Water Security for Resilient Agriculture in India

Sanjay Deshmukh[1]

Hydrological Imperatives and Technological Paradigms in Indian Agriculture

In the relentless pursuit of water-resilient agriculture in the Indian context, comprehending hydrological imperatives and embracing technological paradigms becomes imperative. This component serves as a scientific exploration into the intricate dance between water management, technological advancements and the nuanced implications of genetically modified (GM) crops on water usage. Drawing upon the collective wisdom of agricultural visionaries, this segment unfolds a narrative of sustainable progress and innovation.

Assessing Technological Advancements in Water Management

In the vast tapestry of Indian agriculture, technological advancements have emerged as the linchpin in the quest for sustainable water management. As we meticulously assess the current technological landscape, precision irrigation systems stand out as a beacon of hope, illuminating a path toward water-optimized agriculture. In the arid regions of Rajasthan, for instance, farmers have adopted sensor-based monitoring and automated irrigation technologies, revolutionizing the way water is distributed across fields.

By integrating cutting-edge technologies, such as smart water meters and real-time monitoring systems, Indian farmers gain unprecedented control over their irrigation practices. These tools provide granular insights into soil moisture levels and crop water requirements, empowering farmers to make informed decisions. This precision not only maximizes crop yield but also minimizes water wastage, addressing the critical issue of over-irrigation. Such technological interventions, inspired by the principles of

[1] Professor of Life Sciences, formerly- Vice-Chancellor, University of Mumbai, Mumbai-400098 (MS).

precision agriculture, signify a paradigm shift in the sustainable use of water resources.

To illustrate, in the rice fields of Tamil Nadu, the adoption of precision irrigation techniques has led to a significant reduction in water consumption while maintaining optimal yields. This success story underscores the transformative potential of technology when harmonized with ecological imperatives.

Precision Agriculture Techniques for Water Optimization

The paradigm shift towards precision agriculture marks a pivotal chapter in the ongoing narrative of water-secure farming. This holistic approach, intertwining technology, data analytics and agronomic practices, unfolds a new era in resource optimization, with water conservation at its core. Across the fertile plains of Punjab, satellite imagery. drones and ground-based sensors have become indispensable tools in the arsenal of farmers practicing precision agriculture.

This marriage of data-driven insights and automated machinery allows farmers to tailor their practices to the specific needs of each plot. Variable rate irrigation, guided by real-time data, minimizes water wastage by precisely addressing soil variations and crop conditions. The result is not only enhanced efficiency but also the establishment of a sustainable foundation for water use.

In Karnataka's vineyards, precision agriculture techniques have been employed to optimize water application for grape cultivation. By precisely adjusting irrigation levels based on the specific needs of different grape varieties, farmers have achieved a delicate balance between water conservation and grape quality. This exemplifies how precision agriculture, when integrated into diverse agroecosystems, becomes a powerful tool for ensuring water resilience.

Implications of Genetically Modified Crops on Water Usage

Genetically modified (GM) crops emerge as a compelling dimension in the discourse on water-resilient agriculture. In the labyrinth of India's agricultural landscape, GM crops present a nuanced tapestry of potential benefits and challenges, necessitating a judicious examination of their implications, as underscored by the pragmatic approach of India's agricultural scientists.

Drought-tolerant GM crops, a testament to innovative genetic engineering, hold promise in regions having erratic rainfall patterns. The cultivation of such crops not only ensures food security but also contributes to water-use efficiency by thriving under water-stressed conditions. In Maharashtra's rain-fed regions, where water scarcity is a perennial challenge, farmers have embraced drought-tolerant GM crops, witnessing a significant improvement in crop resilience.

Furthermore, GM crops engineered for pest resistance play a dual role in the water resilience narrative. By reducing the reliance on chemical pesticides, these crops contribute to maintaining water quality and curbing environmental contamination. In the cotton fields of Gujarat, the adoption of pest-resistant GM cotton has not only led to higher yields but has also mitigated the environmental impact of pesticide runoff into water bodies.

However, the deployment of GM crops requires a nuanced approach that carefully navigates ecological, economic and social dimensions. Striking a balance between the potential benefits of GM crops and the need for biodiversity conservation is essential. Lessons from the adoption of Bt cotton in various states underscore the importance of thorough risk assessments and farmer education programmes to ensure responsible deployment of GM technologies.

In conclusion, this segment unfolds a scientific narrative that seamlessly weaves together technological progress, precision agriculture and the role of genetically modified crops in the quest for water resilience in Indian agriculture. This journey, guided by the principles of sustainable practices echoes through the success stories from diverse corners of India. As we navigate this intricate scientific landscape, the pursuit of sustainable practices remains at the forefront, ensuring that technological innovations align with the broader goals of ecological conservation and inclusive agricultural growth.

Strategic Prospects for Water-Secure Agriculture in India

In the labyrinth of challenges surrounding water security for agriculture in India, a profound exploration into strategic prospects becomes imperative. This component intricately weaves the tapestry of population dynamics, evolving dietary patterns and the dynamic demands for agricultural products within a water-centric paradigm. The scientific discourse delves into the complexities of this interplay while illuminating the opportunities that emerge at the nexus of water security and agricultural productivity.

Population Dynamics and Evolving Dietary Patterns

The canvas of Indian agriculture is an ever-evolving masterpiece, shaped profoundly by the dynamics of its population. As we witness transformative shifts in demography, with a burgeoning population and changing dietary preferences, the demand for food surges, ushering in an era that demands a meticulous examination of its implications on water resources. Traditional staples are no longer the sole focus; a discernible shift towards a more diversified and protein-centric diet is evident.

The exponential growth of our population amplifies the quest for food security, exerting considerable pressure on water resources. A holistic solution demands an integrated approach that not only addresses the imperative for increased food production but also recognizes the environmental constraints imposed by water scarcity. This necessitates strategic investments in research and development, coupled with the adoption of innovative farming practices that seamlessly align with the evolving dietary preferences of a dynamic population.

To illustrate, in the fertile plains of Punjab, where the Green Revolution once transformed the agricultural landscape, the population dynamics have catalysed a shift towards high-value crops. Farmers, guided by market demand and changing consumer preferences, are increasingly diversifying into crops such as fruits and vegetables, altering the traditional grain-centric focus. This transition not only addresses dietary shifts but also presents an opportunity for sustainable water use through crop diversification.

Varied Demands for Agricultural Products—A Water-Centric Perspective

A water-centric perspective on the varied demands for agricultural products necessitates a nuanced understanding of the intricate relationships between different crops and their water footprints. It is crucial to recognize that not all agricultural products impact water resources in the same manner and the dynamics of demand must be assessed through the lens of water efficiency.

Certain crops, often labelled as water-intensive, may pose challenges in regions grappling with water scarcity. The delicate balance lies in promoting crops with lower water requirements to ensure sustainability. Embracing climate-resilient and drought-tolerant varieties further mitigates risks associated with water scarcity, aligning agricultural production with the overarching goal of water security.

In the arid landscapes of Rajasthan, where water scarcity is a perennial challenge, farmers have embraced the cultivation of millets. These climate-resilient crops not only require less water but also contribute to diversifying diets and enhancing nutritional security. This exemplifies the symbiotic relationship between varied crop demands, water efficiency and sustainable agricultural practices.

Opportunities for Farmers in The Nexus of Water Security and Agricultural Productivity

The nexus of water security and agricultural productivity unfolds a realm of opportunities for farmers, acting as a beacon of hope in the face of formidable challenges. By embracing water-smart practices, farmers can not only enhance their resilience to climate variability but also contribute significantly to sustainable water use. The integration of efficient irrigation systems, rainwater harvesting and soil moisture management techniques presents promising avenues for optimizing water resources.

Moreover, the exploration of alternative crops that are well-suited to local agro-climatic conditions can diversify income streams for farmers. This diversification, coupled with access to markets for water-efficient produce, contributes to economic sustainability for farming communities. In the rain-fed regions of Andhra Pradesh, the adoption of precision agriculture techniques, coupled with the cultivation of climate-resilient crops like pulses and oilseeds, has transformed the agricultural landscape. This paradigm shift not only enhances water-use efficiency but also elevates farmers' incomes.

The paradigm shifts towards precision agriculture, guided by technology for optimal resource use, further empowers farmers in their quest for water-secure and productive agricultural systems. In the wheat fields of Haryana, for instance, the adoption of precision agriculture practices has led to a significant reduction in water usage while maintaining optimal yields. This success story underscores the transformative potential of precision agriculture when harmonized with the principles of sustainable water management.

In essence, this section underscores the need for a holistic and water-centric approach in shaping the strategic prospects for agriculture in India. By aligning population dynamics, dietary patterns and agricultural demands with sustainable water management practices, we pave the way for a future where agriculture not only meets the growing food requirements but does so in harmony with the finite water resources available. This strategic alignment not only secures the future of agriculture but also ensures the well-being of farming communities and the resilience of the broader ecosystem.

Confronting Challenges in Agricultural Water Security

Agricultural water security in India unfolds at the crossroads of myriad challenges, necessitating a multidimensional approach to unravel uncertainties induced by climate change, mitigate risks arising from erratic weather patterns, address water scarcity challenges, bridge gaps in access to credit and technology and navigate market volatility through holistic water management approaches. In this exploration, it is proposed to delve into the scientific intricacies of each facet, drawing insights from the Indian context and the invaluable experiences of experts in the field.

Unravelling Climate Change-Induced Uncertainties in Water Availability

Climate change stands as a formidable force reshaping the hydrological landscape of Indian agriculture. Rising temperatures and unpredictable precipitation patterns challenge the traditional understanding of water availability. Unravelling the intricacies of climate change-induced uncertainties demands a comprehensive understanding of how these alterations impact regional hydrology.

Aligning with the findings of the Farmers' Commission Report, robust climate modeling emerges as imperative to predict future water availability scenarios. Advanced meteorological tools and satellite-based observations play pivotal roles in anticipating shifts in rainfall patterns, temperature extremes and overall climatic variability. This predictive capability becomes a cornerstone for developing adaptive strategies to ensure water security in agriculture amidst a changing climate.

In the rice fields of West Bengal, where climate change-induced uncertainties in rainfall patterns have become more pronounced, innovative water management practices have been implemented. System of Rice Intensification (SRI) techniques, supported by climate-resilient crop varieties, showcases a holistic approach to unravelling climate-induced water uncertainties. By aligning planting schedules with anticipated rainfall, farmers optimize water use, ensuring sustainable yields even in the face of climate unpredictability.

Mitigating Risks Arising from Erratic Weather Patterns

Erratic weather patterns pose a direct threat to agricultural productivity and water security. Unseasonal rains, unexpected droughts and extreme weather events disrupt traditional farming calendars, necessitating swift adaptation. Drawing from the insights of agricultural visionaries, a resilient agricultural system necessitates adaptive strategies that mitigate risks associated with erratic weather patterns.

The integration of precision weather forecasting, coupled with early warning systems, emerges as a key component in risk mitigation. Providing farmers with accurate and timely information empowers them to make informed decisions regarding planting, irrigation and harvesting. Additionally, resilient crop varieties developed through advanced breeding techniques contribute to mitigating the adverse impacts of erratic weather on crop yields.

In the rain-fed regions of Maharashtra, where unpredictable rainfall patterns have become the norm, community-led early warning systems have proven effective. By leveraging indigenous knowledge and modern meteorological tools, these systems enable farmers to anticipate weather fluctuations and adjust their agricultural practices accordingly. This collaborative approach exemplifies the resilience that can be fostered through the amalgamation of traditional wisdom and scientific advancements.

Water Scarcity Challenges and Their Pervasive Impact on Crop Yields and Economic Outcomes

Water scarcity is a pervasive challenge affecting numerous regions in India. The intricate link between water scarcity and its impact on crop yields and economic outcomes demands a holistic understanding. Sustainable water management and resource-efficient farming practices serve as guiding principles in this context.

Implementing efficient irrigation technologies, such as drip and sprinkler systems, minimizes water wastage and optimizes crop water use. Moreover, the adoption of drought-tolerant crop varieties enhances the resilience of agriculture in water-scarce regions. The economic dimension comes to the forefront as these strategies not only secure crop yields but also contribute to the overall prosperity of farming communities.

In the arid landscapes of Gujarat, where water scarcity has been a perennial challenge, the adoption of conservation agriculture practices has yielded positive outcomes. By integrating cover cropping, minimal soil disturbance and diversified crop rotations, farmers enhance water retention in the soil and reduce dependency on irrigation. This transformative approach not only addresses water scarcity challenges but also fosters economic sustainability by diversifying income streams.

Bridging Gaps in Access to Credit and Technological Resources Amid Water-Related Challenges

Access to credit and technological resources remains a critical bottleneck in the quest for water-resilient agriculture. Several agriculture experts underscore the importance of addressing these gaps to empower farmers. Innovative financial models, such as micro-credit initiatives, coupled with technology transfer programs, become instrumental in abridging these divides.

Farmers' access to modern irrigation technologies, climate-smart practices and precision agriculture tools enhances their adaptive capacity. The integration of digital platforms for knowledge dissemination further democratizes access to information, ensuring that water-related challenges do not disproportionately affect marginalized farming communities.

In the rain-shadow regions of Tamil Nadu, where smallholder farmers often face challenges in accessing credit for agricultural inputs, community-based microfinance initiatives have demonstrated success. By pooling resources and leveraging collective credit, farmers gain access to modern irrigation equipment and climate-resilient seeds. This grassroots approach not only addresses credit gaps but also fosters a sense of community resilience.

Navigating Market Volatility Through Holistic Water Management Approaches

Market volatility amplifies the challenges faced by farmers in water-stressed regions. Sustainable market linkages and integrated water management converge in the pursuit of navigating market uncertainties. Holistic water management approaches involve not only efficient on-farm water use but also strategic planning for market-oriented agriculture.

Water-efficient crop choices, coupled with market intelligence systems, enable farmers to align their produce with consumer demands. This integration ensures that water-resilient agriculture transcends mere productivity gains, contributing to the economic stability of farming communities in the face of market fluctuations.

In the fertile plains of Punjab, where market dynamics often influence agricultural practices, the adoption of crop diversification strategies has proven beneficial. By cultivating high-value crops aligned with market demands, farmers not only optimize water use but also enhance their economic returns. This strategic approach serves as a model for other regions grappling with both water scarcity and market uncertainties.

In conclusion, this segment illuminates the scientific underpinnings required to confront the multifaceted challenges in agricultural water security. By drawing on the wisdom of agricultural experts, it becomes evident that unravelling uncertainties, mitigating risks, addressing water scarcity, bridging access gaps and navigating market volatility demand an integrated and adaptive approach. This scientific discourse lays the foundation for actionable strategies to forge a path towards a water-secure, inclusive and resilient agricultural future in India.

Pathways to Sustainable Livelihoods Through Water-Resilient Agriculture

In the relentless pursuit of water-resilient agriculture, this segment unfurls a comprehensive roadmap that transcends mere adaptation, aiming for transformative shifts in the agricultural landscape. Rooted in an ecological perspective, we need to explore high-value crop diversification strategies, the adoption of agro-ecological practices, harnessing technological innovations for efficient water use and market access, empowering agricultural communities through education and access to credit facilities and building the capacity to thrive amidst the evolving dynamics of agricultural water security.

Transformative Shifts in the Agricultural Landscape—An Ecological Perspective

Sustainable and water-resilient agriculture finds its essence in transformative shifts within the agricultural landscape. Driven by an ecological perspective, this shift surpasses the conventional understanding of agriculture as a production system, envisioning it as an integral part of the broader ecosystem.

The ecological paradigm advocates for the integration of agro-ecological principles, emphasizing the significance of biodiversity, soil health and ecosystem services. The adoption of conservation agriculture, intercropping and agroforestry not only enhances water efficiency but also fosters ecological resilience. This transformative approach seeks to harmonize agricultural practices with nature, ensuring that farming becomes a regenerative force rather than a depleting one.

In the fertile plains of Haryana, where traditional farming practices have often led to soil degradation and water depletion, the adoption of agroforestry has showcased transformative shifts. By integrating fruit-bearing trees with traditional crops, farmers enhance biodiversity, improve soil structure and mitigate water runoff. This holistic approach not only rejuvenates the land but also provides an additional source of income for farmers.

High-Value Crop Diversification Strategies in the Context of Water Security

Diversifying crop portfolios emerges as a strategic imperative for water-resilient agriculture. Recognizing this, the agricultural landscape must pivot towards high-value crop diversification strategies. Dr. Swaminathan's vision for empowering farmers to choose crops based on market demand converges with present advocacy for crop diversification as a risk mitigation strategy.

The identification of high-value crops suited to local agro-climatic conditions becomes paramount. These crops not only offer economic benefits but, when chosen wisely, can contribute to water efficiency. Sustainable agricultural practices, such as precision irrigation and nutrient management, complement crop diversification, ensuring optimal resource use and bolstering the economic sustainability of farming communities.

In the orchards of Himachal Pradesh, where water resources are precious and traditional crops face uncertainties, the cultivation of high-value apple varieties exemplifies the success of diversification strategies. By aligning crop choices with market demand and agro-climatic conditions, farmers not only optimize water use but also secure stable incomes, contributing to the overall economic prosperity of the region.

Adoption of Agro-Ecological Practices to Enhance Water Efficiency

Agro-ecological practices stand as a cornerstone in enhancing water efficiency and resilience in agriculture. Drawing inspiration from the principles laid out various reports brought out by Government of India, agro-ecology aligns agricultural systems with natural processes. Techniques such as rainwater harvesting, cover cropping and integrated pest management not only promote water efficiency but also contribute to soil health and biodiversity.

The adoption of agro-ecological practices requires a paradigm shift in agricultural extension services. Farmers need access to knowledge and resources that empower them to embrace these sustainable practices. Collaborative efforts involving research institutions, governmental agencies and non-governmental organizations are essential in disseminating these practices and ensuring their effective adoption at the grassroots level.

In the rice fields of Kerala, where water scarcity poses challenges, the implementation of System of Rice Intensification (SRI) practices showcases the effectiveness of agro-ecological approaches. By optimizing planting densities, improving soil structure and reducing water use, farmers not only enhance water efficiency but also witness increased yields. This sustainable model exemplifies the potential of agro-ecological practices in addressing water challenges.

Harnessing Technological Innovations for Efficient Water Use and Market Access

Technological innovations play a pivotal role in shaping the future of water-resilient agriculture, where harnessing of technology for efficient water use and improved market access becomes instrumental in transforming the agricultural landscape.

In the tech-savvy farms of Karnataka, where water management is crucial, the adoption of sensor-based irrigation systems has revolutionized agricultural practices. By precisely monitoring soil moisture levels and automating irrigation schedules, farmers optimize water use, reduce wastage and enhance crop yields. This technological intervention not only ensures efficient water utilization but also contributes to the economic upliftment of farmers through increased productivity.

Empowering Agricultural Communities Through Education and Access to Credit Facilities

Empowering agricultural communities is central to the sustainable transformation of the sector. Commitment to inclusive growth and recognition of the importance of farmer education need to be converged in the emphasis on empowering farming communities through education and access to credit facilities.

Education becomes a catalyst for change, enabling farmers to adopt modern agricultural practices, make informed decisions and adapt to evolving challenges. Simultaneously, access to credit facilities ensures that farmers have the financial means to implement sustainable water management practices, invest in technology and weather unforeseen challenges. A robust extension system that combines education and credit access becomes integral to building the capacity of farmers to thrive in a changing agricultural milieu.

In the villages of Uttar Pradesh, where educational resources are sometimes scarce, community-led initiatives have successfully empowered farmers through education and credit access. Farmer study groups, supported by local agricultural extension services, provide a platform for knowledge exchange and skill-building. Concurrently, micro-credit initiatives enable farmers to invest in water-efficient technologies, contributing to both ecological sustainability and economic prosperity.

Building Capacity to Thrive in the Evolving Dynamics of Agricultural Water Security

Building capacity to thrive in the evolving dynamics of agricultural water security requires a holistic approach that encompasses knowledge, skills and resilience. Here, call for capacity-building and acknowledgment of the importance of farmer training need to be converged in the recognition that an empowered and resilient farming community is essential for sustainable agriculture.

Capacity-building initiatives should focus on enhancing farmers' understanding of water-smart practices, technology adoption and market-oriented agriculture. Training programs, demonstration farms and farmer field schools serve as effective platforms for knowledge dissemination and skill development. This, coupled with the provision of extension services, ensures that farmers are well-equipped to navigate the complexities of agricultural water security.

In the rain-fed regions of Assam, where climatic uncertainties are prevalent, capacity-building initiatives have bolstered farmers' resilience. Farmer field schools, organized by local agricultural universities, impart practical knowledge on water-saving techniques, crop diversification and market linkages. This grassroots approach not only enhances farmers' capabilities but also fosters a culture of continuous learning and adaptation.

This segment delineates a scientifically validated blueprint for pathways to sustainable livelihoods through water-resilient agriculture. By embracing transformative shifts, high-value crop diversification, agro-ecological practices, technological innovations, education and capacity-building, this component charts a course toward an agricultural future that is not only resilient to water challenges but also ensures the prosperity and well-being of farming communities in India.

A Call to Action for a Water-Secure, Inclusive and Resilient Agricultural Future

In the current epoch of India's agricultural narrative, this narrative stands as a profound call to action, resonating with urgency and purpose. This section delineates a meticulous blueprint inspired by visionary leadership and pragmatic insights, aiming to foster sustainable water management, implement strategic interventions for addressing water security challenges, institute policy reforms harmonizing water conservation and agricultural development, nurture community-based initiatives for water-resilient agriculture and unequivocally prioritize the well-being and prosperity of agricultural communities. This scientific exposition refrains from overt attribution but draws wisdom from the collective ethos of leaders in agricultural science.

Concerted Efforts for Sustainable Water Management in Agriculture

At the core of a water-secure agricultural paradigm lies the imperative for concerted efforts towards sustainable water management. The confluence of advocacy for integrated water resources management and emphasis on efficient water use underscores the urgency of this imperative. Sustainable water management necessitates the assimilation of cutting-edge technologies encompassing irrigation, soil moisture monitoring and water-use efficiency.

Best practices such as rainwater harvesting, watershed management and the promotion of water-saving irrigation techniques take precedence in this comprehensive strategy. In the arid landscapes of Rajasthan, where water scarcity has posed perennial challenges, the deployment of precision irrigation technologies has emerged as a transformative force. By precisely calibrating water delivery to crops as per their needs, these (precision irrigation technologies) technologies not only conserve water but also amplify agricultural productivity, providing a beacon of hope for water-scarce regions nationwide.

Strategic Interventions to Address Water Security Challenges

Addressing the intricate tapestry of water security challenges demands strategic interventions rooted in scientific rigor and practical wisdom. The synergy between insistence on adaptive strategies and pragmatic resilience in the face of challenges accentuates the necessity for interventions addressing the multifaceted dimensions of water security. Early warning systems for water scarcity, water-use efficiency benchmarks for agriculture and a dynamic policy framework responsive to evolving climatic conditions constitute imperatives in this context.

Investments in research and development targeting water-efficient crop varieties and technologies play a pivotal role in bolstering agriculture against the uncertainties posed by water scarcity. In Maharashtra, characterized by water-stressed terrains, strategic interventions such as the introduction of drought-resistant crop varieties and adoption of efficient water storage practices have proven instrumental. These interventions not only mitigate the impact of water scarcity but also furnish a blueprint for sustainable agricultural practices nationwide.

Policy Reforms for Integrating Water Conservation and Agricultural Development

The nexus between water conservation and agricultural development emerges as a pivotal focal point for strategic policy reforms. The convergence of policy recommendations for sustainable agriculture with advocacy for policy frameworks stimulating agricultural productivity underscores the need for reforms seamlessly integrating water conservation into agricultural development. Incentivizing water-smart agricultural practices through subsidies, research funding and market-oriented incentives assumes critical significance.

Zoning regulations and land-use policies should harmonize with water availability, ensuring agricultural activities align with the carrying capacity of the region. Policies facilitating the integration of traditional wisdom with modern technologies establish a holistic approach addressing both conservation and development imperatives. In water-abundant regions such as Punjab, policy reforms encouraging the adoption of conservation agriculture practices have heralded improved soil health, reduced water runoff and heightened water-use efficiency, setting benchmarks for nationwide emulation.

Community-Based Initiatives to Foster Water-Resilient Agriculture

The empowerment of local communities constitutes a pivotal strategy in fostering water-resilient agriculture. The alignment of emphasis on grassroots participation with the design of community-based initiatives harnessing local knowledge, promoting collective water management and empowering farmers as stewards of water resources signifies an essential paradigm shift. Farmer cooperatives for water-sharing, community-led water conservation projects and the establishment of local task forces for climate-smart agriculture emerge as indispensable components.

Implementation of such initiatives not only fosters sustainability but also instils a sense of ownership among farmers towards water management practices. In the rain-fed regions of Odisha, community-led watershed management projects stand as exemplars in the rejuvenation of water resources and the improvement of agricultural productivity. These initiatives underscore the latent potential of community-based approaches in achieving water-resilient agriculture, thereby reinforcing the relevance of decentralized strategies in ensuring sustainable water utilization.

Prioritizing the Well-Being and Prosperity of Agricultural Communities in the Water Security Agenda

The overarching goal of a water-secure, inclusive and resilient agricultural future remains incomplete without an unwavering commitment to prioritizing the well-being and prosperity of agricultural communities. The alignment of emphasis on inclusive growth with the recognition of the social dimensions of agricultural development accentuates the paramount importance of this principle. Prioritizing the well-being of agricultural communities entails the creation of social safety nets, ensuring access to healthcare and education and fostering gender equity in agricultural activities.

Economic prosperity is measured not merely in terms of increased productivity but also in the enhancement of the quality of life for farming communities. In the tribal regions of Jharkhand, holistic initiatives encompassing education, healthcare and sustainable livelihoods have wrought transformative changes in the lives of farming communities. These examples underscore the holistic approach essential to ensuring the well-being and prosperity of those who constitute the backbone of our agricultural landscape.

This segment serves as a rallying point for transformative action, inspired by visionary leadership and grounded in pragmatic insights. By intricately weaving together sustainable water management, strategic interventions, policy reforms, community-based initiatives and an unwavering commitment to the well-being of agricultural communities, this call to action charts a definitive course towards a water-secure, inclusive and resilient agricultural future for India.

Water Security and Novel Solutions for Agriculture

Ashwin Shroff[1]

Sensitivity to environmental issues has been increasing by the day. We have been witnessing effects of global warming and increased pollution of air, water and environment. This change in environment significantly affects agriculture and allied activities. Agriculture has become vulnerable and risky profession. While water is a basic need for agriculture and needs better management through recycling waste water and water harvesting, other basic elements are soil and weather. Use of IT and other technologies has also become imperative.

Green revolution gave us food security. It mainly depended on good seeds, water and chemical fertilizers and pesticides. However, now, the situation has changed very much and the context has changed.

Excel group has been actively engaged with Agriculture, farmers and rural communities. There are several initiatives, many for the first time in India and even globally e.g. Chemical pesticides, drip systems. Compost from organic wastes, sewage water treatment. We continue to be active through activities like water security & availability, climate change reversal through seaweeds and soft chemicals and biotech products for sustainable agriculture. Some of our solutions are as follows.

Waste water treatment through Biofilter: Treatment of water using simple, decentralized and environment friendly methods using minimal energy. No sludge is generated in the process as all solids get converted into compost.

Hydroponic systems for fodder grass cultivation: Fodder grown is high in nutrient compare to traditional green fodder and requires less water and is also fast growing.

Water Conditioner and pH manager: It prevents interaction between pesticides or water-soluble fertilizers and dissolved solids in water. It thus prevents precipitation of water-soluble fertilizers and pesticides, improving its performance. This implies reduced consumption of pesticides or fertilizers and saving to farmer.

Hardness of water: We have developed solution which reduces the hardness of water by treating the irrigation water with high TDS and thus

---

[1] Chairman, Board of Directors, Excel Industries, Mumbai.

augmenting the nutrient carrying capacity of the irrigated water at par with the rain water.

De-clogging of drip system: Many fields do have drip systems these days. Fertilizers and other agro inputs are delivered through the drip systems. These drip systems get clogged with Ca/ Mg salts as well as fertilizers. Farmer has to spend for removal of lateral and then wash it in to a tank containing mineral acids. This implies playing with hazardous acids besides extra expenditure on labour to remove and lay back the lateral pipes. We have developed soft solution which has the de-clogging or anti-scaling properties and when applied in the drip systems, it de-clogs the dripper.

Bio stimulant: A large portion of nutrients gets fixed in soil. Our bio stimulant product releases these nutrients and make them available to the crop and thus enhancing the productivity.

Composting Culture: We have developed a culture which works in aerobic manner to decompose the organic waste material like agri waste and municipal solid waste to usable compost.

Our associated NGOs have been actively involved with several agriculture and allied activities.

- Water harvesting and irrigation.
- Animal husbandry and dairy.
- Agriculture development training programs.
- Weather forecasting and climate management solutions and providing weather related information to farmers.
- Drought proofing programmes.
- Agri extension services.
- Besides basic needs of agriculture, agriculture in future will require precision farming and will use IT and other technologies.
- Precise application of inputs at the right place, at the right time and in the right amount to minimize the production cost, increase profitability and reduce risks.
- Data driven farming for analysing and correlating information about weather, types of seeds, soil quality, probability of diseases. Every field to be managed as per site specific data and information.
- Use of drone for soil health assessment, crop monitoring and sensor-based drone for irrigation.
- Small automated weather station for getting information about a particular small area for early warning. e.g. Temperature, humidity, wind speed, wind direction, likely precipitation.

Impact of Surya Shakti Kisan Yojana on Water Markets in Gujarat*

Sonal Bhatt[1]

Background

Ground-water is a free, non-divisible natural resource devoid of an economic value (price). However, when only few farmers own infrastructure for groundwater withdrawal such as wells, tube wells, irrigation pumps (IPs) through which they could sell water to other farmers; groundwater becomes a shareable resource attached with an economic value in the form of its price, giving rise to groundwater markets (Manjunatha *et al.*, 2011). The price of groundwater is determined by various factors. Apart from policy interventions, many other factors such as hydro-geological characteristics, fixed costs of drilling wells; path dependence in the construction of irrigation infrastructure (wells and pipelines) (Dubash, 2000), price of power, hours of power supply, distance between the source of water and its point of use; non-economic factors such as historical and social norms as well as local institutions, social relations based on caste and kinship etc. Singh (2007) noted that the system of groundwater exchange in Gujarat does not completely follow the market forces of supply, demand and competition. Rather, it is an 'instituted process' in the context of historical, social as well as hydrological factors. Tamuli *et al.* (2021) observed that tube well-specific factors such as excess capacity of the pump sets, farm-specific factors such as availability of diesel operated pump sets and conveyance facilities, depth of the borewell etc. as well as social factors such as the caste of the sellers and buyers, were the significant in determining the water selling decisions of farmers.

As water becomes a tradable commodity, farmers adopt water-saving technologies and switch to high-value crops that are less water-intensive. Shah and Chowdhury (2017) noted that India is the world's largest

---

* This research was conducted as a part of a Consultancy assignment completed for the International Water Management, Colombo in 2023.

[1] Professor, Department of Economics, Sardar Patel University, Vallabh Vidyanagar, Gujarat.

groundwater irrigator with a majority of marginal as well as tenant farmers who are heavily dependent on informal water markets for their irrigation needs. Besides, water markets are not only beneficial in water scarce regions but also in water surplus regions where they could transform a stagnant agricultural economy (Shah and Ballabh, 1997) into a vibrant one, with positive impacts on expansion of access to irrigation (Manjunath *et al.*, 2011); increase in marginal productivity, equity of income distribution and expansion of labour use in agriculture; halting a decrease in groundwater tables; improved cropping patterns of water buyer creating in turn, more opportunities for employment (Shah and Ballabh, 1997).

Kolvalli and Chocoine (2007) observed that sale of groundwater was also source of income for the water sellers that incentivized them to make investments in irrigation infrastructure. Singh (2007) revealed how the tube well owners enjoyed a monopoly and aimed at profit maximization through water sale. Since water sale was possible only with certain geographical limitations, it sometimes also resulted in a situation of monopsony of the buyers. However, both these situations were harmful in terms of resource sustainability.

Shah and Chowdhury (2017), Kumar and Singh (2001) and Dubash (2002) raised concerns about the provision of subsidies on electricity that has made groundwater withdrawal for irrigation a very cheap proposition, which serves to encourage the tube well owners to sell more groundwater. However, this seemingly 'pro-poor' scenario consisting of cheap electricity-led expansion of the groundwater markets may have potentially disastrous ecological and social consequences for groundwater as a resource. It results in too much withdrawal of groundwater which in turn results in a depletion of groundwater tables on one hand and immense burden on the finances of electricity DISCOMs (distribution companies) on the other hand. Shah and Ballabh (1997) concluded that the efficiency of the power supply regime was a major barrier against the expansion of equitable water markets; while the price and availability of diesel for pumping was also a major factor of influence.

Worldwide, the off-grid solar PV irrigation pumps are expected to expand the access to groundwater and reduce the cost of energy and the carbon emissions during its production. This is the way forward towards combating climate change. In developing countries across the Middle East and North Africa, South-east Asia and Sub-Saharan Africa, the adoption of solar technology in agriculture to lift groundwater is rapidly expanding, following decreases in pump costs, economic incentives and development partner initiatives. Solar irrigation potentially provides a cost-effective and sustainable energy source to secure food production and sustain livelihoods in line with multiple Sustainable Development Goals. Lefore et al. (2021) propose a framework to support policy, regulation and monitoring for environmentally sustainable and socio-economically inclusive solar irrigation for smallholder agriculture. Pervasive policies on electricity supply and price in India have resulted in uncontrolled groundwater withdrawal and threatened to affect climate change through carbon emissions resulting from the generation of electricity, besides leading to an overuse of groundwater resources for irrigation and the depletion of aquifers. This has led to more and more energy consumption and carbon emissions, as more energy is required to pump water from deeper wells. Solar energy-based irrigation pumps could be a panacea to this problem by reducing the pressure of electricity supply on the grid and at the same time, reducing the carbon emissions generated during the production of electricity. Solar irrigation pumps in India could replace diesel-powered pumps in many parts of India's agricultural heartlands, especially in the under-electrified Gangetic plains. They could even increase agricultural output by making more irrigation water available to farmers (https://bridgetoindia.com/the-market-for-solar-irrigation-pumps-in-india/).

Closas and Rap (2017) highlighted the need to further study the impacts, opportunities and limitations of solar photovoltaic (PV) pumping systems in groundwater-fed agriculture. They showed how most policies and projects promoting solar-based groundwater pumping for irrigation through subsidies and other incentives overlook the real financial and economic costs of this solution as well as the availability of water resources and the potential negative impacts on the environment caused by groundwater over-abstraction. They opined that there is a need to monitor groundwater abstraction, targeting subsidies and improving the knowledge and monitoring of resource use while promoting solar PV based pumping systems for irrigation. Failing to address these issues could lead to further groundwater depletion, which could threaten the sustainability of this technology and dependent livelihoods in the future.

Water markets in Gujarat have a long history; and norms of water buying and selling, water prices, terms and conditions of water sale are well-established; even though informal in nature. Degradation and depletion of groundwater resources is emerging as a major concern in many arid and hard rock sections of Gujarat (Dubash, 2002); and more so because it disproportionately affecting poor farmers. Equitable access to groundwater is as important as its sustainable use.

Water rates and terms of water sale have been traditionally influenced by the prices and supply situation of electricity from time to time. Shah et al., (2008), observed that in Gujarat, groundwater over-exploitation since 1980s was the result of huge energy subsidies that encouraged groundwater overdraft. It resulted in a bankrupt electricity DISCOM, i.e. Gujarat Urja Vikas Nigam Ltd. (GUVNL) on one hand and depleted aquifers on the other. Government of Gujarat (GoG) has undertaken various measures for the restriction on electrification of tube wells in over exploited, dark and saline areas in order to regulate the rate of groundwater withdrawal and depletion. The various electricity supply and pricing regimes prevalent in Gujarat from time to time are discussed as follows:

Electricity Supply and Price regimes in Gujarat

Phase I

Irrigation pumps were started to be electrified in Gujarat in the 1970s. Initially, electricity tariff was charged at a flat rate per year. Marginal tariff was constant irrespective of actual consumption of electricity, i.e. irrespective of the amount of water pumped. This gave an impetus to incessant withdrawal of groundwater; which was, even otherwise a free natural resource. This policy continued till about the 2000s.

Phase II

After 2000s, a metered tariff rate was introduced in which, payment for electricity was connected to the amount of its consumption. This was an important step towards rationalizing the use of electricity and in turn, curtailing the withdrawal of groundwater through electrified pumps. While on one hand it could improve the revenue for the DISCOM, on the other hand it could also regulate the amount of groundwater being pumped out. Supply of electricity for both farmers as well as domestic users was from the same feeder. However, agricultural users were provided with three-phase power only for eight hours a day, while domestic users were supplied with single-phase electricity for twenty-four hours a day. Many farmers took undue advantage of this arrangement. They illegally converted single-phase connections into three-phase ones and in turn, exploited not only electricity supply system but also depleted groundwater by withdrawing too much of it. Evidence suggests that water sellers earned a sizeable portion of their incomes from selling water.

Phase III

In 2005–06, the GoG introduced the Jyoti Gram Yojana (JGY). Under JGY, US$ 290 million were invested for separating agricultural feeders from domestic feeders. Agricultural connections were promised an assured uninterrupted 3-phase supply of electricity for 8-hours alternately during the day or night on a fortnightly rotational basis. At the same time, domestic connections were to get a 24-hour uninterrupted single phase power supply. This established a tight regimen for farm power rationing in the rural areas and wrung a noose around the illegal practice of phase conversion. Electricity tariff was Rs. 0.50 per unit, which was later increased to Rs. 0.60 per unit. As power was rationed, water prices charged by tube well owners increased by 30–50 per cent. While it brought a positive effect on groundwater withdrawal by tube well owners; it had a negative impact on the income and livelihoods of marginal and landless farmers, who were water buyers and dependent on tube well owners for their irrigation needs. Nevertheless, JGY was a significant policy measure that albeit indirectly, affected groundwater withdrawal in terms of curtailing the excessive pumping of ground water through illegal means. Besides, an assurance of 8-hour agricultural power supply—albeit for limited hours—led to more efficient use of groundwater (GoI, 2007).

Phase IV

In 2018, the GoG introduced the *Surya Shakti Kisan Yojana* (SKY); operationalized by the state-run electricity DISCOMS (distribution companies) organized under the umbrella of *Gujarat Urja Vikas Nigam* Ltd. Under SKY, the farmers were to install grid-connected solar PV panels in their farms. They could use the solar power generated by them for pumping out groundwater for their own use as well as for selling it to their buyers. The excess power generated by them could be evacuated in the electricity grid belonging to the DISCOM. The sale of this excess power to the DISCOM could bring additional income to them.

Accordingly, solar panels were installed on those farms which already had an electricity connection. At this juncture, the farmers were given a choice regarding the number of panels they wanted to install and the HP of the electricity connection could also be ungraded if they wished to do so. In other words, the existing metered electricity connection was converted into a SKY connection.

In order to meet the initial (high) cost of installation of the solar apparatus, a 60% subsidy was to be given which was to be shared equally by the Central and the State governments. Another 35% of the project cost was to be arranged for the farmer through loan taken from the World Bank at a rate of interest of 6% p.a. by the State government on behalf of the farmers and was meant to be repaid by them with the help of the income earned through the sale of power. The remaining 5% of the project cost was to be paid upfront by the farmer participating in SKY. The total duration of the Scheme is 25 years which was split between 7-year period and 18-year period.

The farmers would be paid Rs. 7 per unit of power (Rs 3.50 by GUVNL + Rs 3.50 by State Govt.) that they sold to the grid for the first 7 years. Out of this, Rs. 3.50 would be collected at source against the repayment of the loan taken from the World Bank. Thus, in effect, the farmer would be paid Rs. 3.50 for each unit of solar power evacuated to the grid for the first 7 years. In the subsequent 18 years, the farmers would get the rate of Rs. 3.50 for each unit of power sold by them. Thus, for a period of 25 years, the financial arrangements of the scheme were binding to both the parties i.e. the state government via the DISCOM as well as the farmer. However, in effect, the contract designed for SKY was a tri-partite one. It is made between three parties viz. the DISCOM, the farmer as well as the empanelled agency responsible for installation and/or maintenance of the apparatus for generating solar power.

The GUVNL had calculated loan repayments in such a way that unless farmers pumped less than 800 units/HP per year, i.e. they withdrew more power from the grid than what they evacuate to it, they would receive a payable amount in their annual electricity bills. Thus, there was an incentive to reduce hours of pumping. Overall, the incentive structure of SKY discouraged farmers from selling water.

Farmers who joined SKY would receive 12 hours of power supply during the day time. This was 1.5 times more than previously (under JGY) when it was available only for 8 hours a day; at either day or night time, on rotational basis.

SKY was an important intervention in the agricultural sector of Gujarat. Shah and Rai (2021) believed that SKY would serve as a politically acceptable solution to the vicious circle of low energy prices-excess-groundwater withdrawal-depletion of groundwater tables-increased demand for electricity which was prevalent in Gujarat. They hoped that SKY could have a significant impact on multiple objectives such as lowering the groundwater withdrawal, reducing the rate of groundwater depletion, reducing the GHG emissions resulting from electricity generation, combat climate change and last but not the least, reducing the huge burden of subsidies on the state exchequer resulting from farm power supply by the DISCOM. They recommended that the GoG should invest in free of cost but compulsory solarization of agricultural tube wells and demonstrated it to be a profitable option not only for the GoG but also for the farmers. The theoretical possibilities of the impact of SKY on informal water markets could be listed out as follows:

1. Solarized irrigation pumps (SIPs) under SKY would use freely available solar power. Besides, groundwater was already free and electricity was now available for 12 hours instead of 8 hours previously. If the tube well owners increased withdrawal of groundwater for their own consumption as well as for sale, this could result in a lowering of water prices for irrigation which would obviously benefit water buyers but could have a potentially disastrous impact on the groundwater tables in the area.
2. SKY was designed in such a way that farmers could earn income only if their net power production was positive, i.e. they consumed less power than what they produced through the solar panels. They could do this either by selling lesser water than before or increasing its price, or both.
3. If sale and purchase of groundwater is prominently affected by non-price factors such as social ties rather than economic factors such as the cost of power; SKY sellers may not find it easy to reduce their water selling. However, as per the design of SKY, if they continue selling the same amount of water, they would become net consumers of electricity instead of net producers; and would be required to pay electricity bills to MGVCL. Hence, they would have to balance their non-economic compulsions against the economic ones.
4. Groundwater withdrawal for irrigation may be inelastic in nature; and in turn, inelastic to the requirement and the price of power. Hence, if the price of solar power offered by the DISCOM is not attractive enough, it may not be able to make a substantial impact on use of power for groundwater withdrawal.

In the light of the above theoretical possibilities, the following objectives were formulated for this study.

1. To understand the implementation of SKY in Anand district.
2. To understand the impact of SKY on the withdrawal of groundwater, groundwater table, its price and the area irrigated with the help of

ground water by various stake-holders such as SKY sellers, non-SKY sellers (water sellers without SKY connection) and water buyers.

3. To understand the importance of non-economic factors viz-a-viz economic factors in the determination of price of groundwater.

This study intends to answer the above by way of a qualitative study done in Anand district in Central Gujarat. Anand district was purposively chosen for this study as it is groundwater rich and has traditionally vibrant agriculture with well-developed water markets. Dhundi village in Kheda district near Anand is the location where the world's first solar irrigation cooperative was established successfully (Bhatt and Kalamkar, 2017).

Three feeders of the DISCOM functioning in Anand district, i.e. *Madhya Gujarat Vij Company Ltd.* (MGVCL), namely Ashapuri, Ishnav and Trambovad were purposively selected from the Petlad division of MGVCL. The selected feeders had several villages under their purview; where in some instances, multiple feeders were servicing the same village while a single feeder was also found to be servicing several villages. The details about the study area are presented in Table 1,

Table 1: Selected Villages and Feeders in Anand District

Sr. No.	*Village*	*Taluka*	*Feeder*	*Geographical Area (in ha.)*	*Population (No. of persons)*	*No. of HHs*	*Literacy Rate (%)*
1.	Rangaipura	Petlad	Ashapuri	246.69	4650	1044	83.46
2.	Palaj	Petlad	Ashapuri	787.49	9781	2112	74.12
3.	Piplav	Petlad	Ishnav	595.73	5000	1056	83.50
4.	Ishnav	Sojitra	Ishnav	456.26	3222	725	66.85
5.	Trambovad	Sojitra	Trambovad	384.72	5453	1164	65.05

Source: https://villageinfo.in/gujarat/anand/sojitra/isnav.html, https://villageinfo.in/gujarat/anand/sojitra/trambovad.html,https://villageinfo.in/gujarat/anand/petlad/palaj.htmlhttps://villageinfo.in/gujarat/anand/sojitra/piplav.html#:~:text=Piplav%20has%20a%20total%20population,piplav%20village%20locality%20is%20388460, https://villageinfo.in/gujarat/anand/petlad/rangaipura.html

Major crops grown include tobacco and paddy in *Kharif*, wheat in *Rabi* and tobacco, pearl millet and wheat in summer. Many farmers also chose to keep their land fallow or grow only fodder crops during summer. The feeder-level details of SKY and non-SKY electricity connections as well as details about the availability of canal irrigation are presented in Table 2 as follows:

Table 2: SKY and Non-SKY Electricity Connections in the Selected Feeders

Name of the Feeder	*Villages*	*Total SKY Connections*	*Total Non-SKY Connections with Watch Dog Transformers (WDTs) Installed*	*Availability of Canal Irrigation*
Ashapuri	Rangaipura Palaj	32	19	River Mahi flows near Rangaipura; hence it is included in Mahi Right Bank Canal (completed in 1958) command area
Ishnav	Piplav Ishnav	82	21	Ishnav included in MRBC command area
Trambovad	Trambovad	83	15	Included in MRBC command area
Total		197	55	

Source: Records provided by the MGVCL, Anand, https://guj-nwrws.gujarat.gov.in/showpage.aspx? contentid=2953 &lang =Gujarati accessed on November, 01, 2022

Sample Selection and Data Collection

Individual water sellers and water buyers were selected from each of the three feeders through purposive convenient sample of SKY sellers and Non-SKY sellers. Sampling was curtailed by the fact that many water sellers were not available on site, were living abroad or out of station, leaving their farms to care-takers or labourers. Hence, they could not be contacted. Similarly, a comprehensive list of water buyers was not available; hence they were contacted using the snowball method with the help of information provided by their water sellers. Qualitative data was collected from i) SKY water sellers, ii) non-SKY (electrified) water sellers iii) water buyers from SKY sellers. The feeder-wise distribution of individual SKY sellers, non-SKY sellers and SKY buyers is presented in Table 4 and Table 5 respectively.

Table 4: Distribution of Selected SKY and Non-SKY Water Sellers

Name of the Feeder	*SKY Water Sellers*	*Non-SKY Water Sellers*	*Total Selected Sellers*
Ashapuri	09	04	13
Ishnav	19	02	21
Trambovad	21	00	21
Total	49 (24.87%)	06 (10.90%)	55 (21.82%)
	N= 197 (100%)	N=55 (100%)	N=252

Source: Personal interviews conducted by the Author.

Table 5: Distribution of Selected Water Buyers from SKY Sellers

Name of the Feeder	*Selected Water Buyers from SKY Sellers*
Ashapuri	06
Ishnav	12
Trambovad	18
Total	**36**

Source: Personal interviews conducted by the Author.

This is an exploratory study of qualitative analysis was conducted between June–July, 2022. Collected data was recorded on perceptions, expressions, opinions and observations made by the respondents in their native language. Thematic analysis of this data was done, which is presented in the form of Tables and diagrams.

Results and Discussion

The tube well owners preferred to sell water not merely to 'generate additional income' but more importantly, to 'help neighbours or extended family members or close relatives' who may be adjacent land holders. It was very important for them to 'maintain social ties'. If they didn't sell water, 'what would the buyers do without it?' Caste and kinship relations were maintained over generations together; so, buying water from or selling water to, became obligatory in many cases regardless of its economic implications. Many tube wells in the study villages were as old as 50 years or more, hence water buyers had been buying water from their tube wells since more than one generation. Stopping or reducing irrigation service to them was quite impossible and left the pump owners did not 'have much choice' or 'didn't want to choose to stop selling water'. Some farmers were obliged to sell water on one land parcel because they were required to buy water from others at another land parcel. Even if they could increase water rates notionally, it was not always possible to collect the same, because, personal relations and social ties came in the way of bargaining hard and negotiating favourable terms of payment.

As presented in Table 6, the respondents found the flat-rate regime of electricity prices to be the 'cheapest' and 'most affordable'. However, it had issues in terms of quality and reliability of power supply. Even though cheap, power was available sparingly and power outages were common. This system was replaced by the metered rate regime in the 2000s. While the price of power was rationalized, it was not perceived to be too high, compared to the cost of diesel-powered irrigation. However, it became very difficult to acquire new connections and one had to go through a long waiting period for acquiring one. Besides, the earlier problems of load-shedding and voltage fluctuations continued unabated.

The JGY was a game-changer in the electricity supply and price regime because it promised 8 hours assured power supply to agriculture through exclusive feeders. Farmers found this system the most beneficial of all systems till date, even though they revealed that even under JGY, power was supplied for lesser than the promised 8 hours and power outages didn't end.

Table 6: Advantages and Disadvantages of Various Electricity Supply and Price Regimes to the Farmers

	Advantages	*Disadvantages*
Flat Rate (1970s onwards)	1. Fixed charge per connection per year 2. 'cheapest' and 'affordable' power 3. In 1993–94, flat rate connections attracted an annual bill of about Rs. 7500 only	1. Frequent power outages 2. Voltage fluctuations 3. Uncertainty of supply
Metered Rate (Introduced in about 2000–01)	Price charged according to units of consumption per year	
Jyoti Gram Yojana (Introduced in 2005–06)	1. Price charged according to units of consumption per year 2. Unit rates reasonable 3. Assured 3 phase electricity supply for 8 hours at day/night on rotational basis at day and night 4. Single phase electricity for	1. Costlier than flat-rate regime 2. Night-time irrigation found 'inconvenient' 3 Even though power supply promised for 8 hours, actually much

	Advantages	*Disadvantages*
	residential connections for 24 hours, so 'benefitted our homes' 5. No. of tube wells and irrigated area increased a lot 6. Beneficial for tobacco crop in which day-time irrigation damaged the crop roots 7. Expenses on electricity lower compared to irrigation by diesel-powered pumps	less and only intermittent 4. Outages occurred sometimes
SKY upon inception (2018–2019)	1. Thought that Solar power would be available for 'free' 2. Thought that farmers could earn additional income if produced more power than their consumption; had to pay if consumed more than production 3. Upfront investment of 5% very 'affordable' 4. Attractive subsidy from State and Central governments 5. Convenience of day-time electricity 6. supply 7. Supplied regularly for 12 hours instead of 8 8. Day-time electricity available from 06:30 am to 06:30 pm; 9. No botheration of irrigating at night 10. Expenses on electricity irrigation expected to be lower	1. Sometimes power evacuation to the grid is halted 2. Day-time irrigation not expected to be favourable for tobacco crop
SKY before payable bills (2019–20)	1. Solar power produced on farms and evacuated in the electricity grid 2. Sufficient irrigation for all including buyers, because of the extended hours of power supply 3. No botheration of irrigating at night	1. Not favourable for tobacco crop 2. No change in irrigated area
SKY after payable bills (2020–21) and 2021–22)		1. Contrary to expectations, electricity use became costlier 2. Unit rates charged for electricity consumed in

	Advantages	*Disadvantages*
		excess of production is 'high' to 'too high' 3. Irrigation became costlier 4. Number of irrigations supplied to the buyers had to be decreased 5. Some water buyers had to be off-loaded 6. Crops requiring lesser irrigation such as castor, mustard, preferred 7. Started taking two seasons instead of three 8. Avoiding summer crop 9. Instead found it more lucrative to sell power during summer

Source: Primary Survey conducted by the Author.

Farmers opted for SKY as they were 'attracted by the substantial subsidy being offered by the state as well as central government'. The upfront investment for getting a SKY connection was only 5% of the total payment, which was quite affordable. Assistance for this was arranged in the form of a loan from their cooperative credit society. On getting a SKY connection, the farmers expected to 'earn an additional income by selling power to the MGVCL', apart from getting '12-hour day-time electricity'. The experience with SKY was mixed; and changed with each passing year. SKY had increased the hours of supply of irrigation and eliminated the need to undertake night shifts for irrigation. However, after receiving payable bills instead of 'free solar power', irrigation became much costlier than expected and it was revealed that 'many farmers are choosing to take only two agricultural seasons per year' and 'keep the land fallow during summer'; choosing to sell only power in the summer months.

Farmers experienced a steady rise in the price of irrigation water over the years. While the flat-rate regime was the cheapest, there was a gradual increase in the water rates during the transition to flat rates to metered rates. Water rates during the flat rate regime were about Rs. 25–30 per hour which doubled to about Rs. 48–60 per hour when agricultural connections were metered; and remained almost unchanged even after the introduction of JGY in 2005. However, during the SKY regime, changes in water rates have been very rapid. While immediately after SKY (when the sellers had not received payable bills), the water rates rose only slightly; in subsequent years they were more than doubled. After SKY farmers started receiving payable bills, they doubled their water rates up to Rs. 80–120 per hour. On the other hand, non-SKY farmers continued to sell water at about the same rates as before.

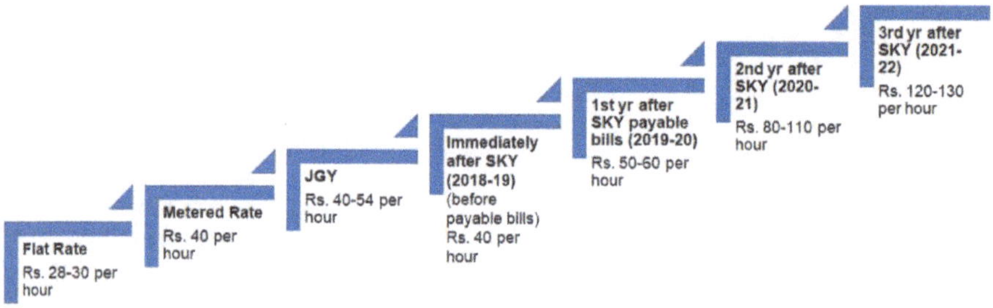

Source: Qualitative data collected by author.

Farmers' Experiences with SKY

(a) *Immediately after SKY (2018–19)*

In the year immediately after the implementation of SKY, water sellers and buyers alike were delighted that their pumps would run on freely available solar energy while the hours of electricity supply were increased to 12 hours during the day-time, bringing more convenience as well. This was the year when they had not yet received any payable electricity bills from MGVCL and made no change in the water rates.

(b) *1st year after SKY payable bills (2019–20)*

Most farmers received payable bills at the end of the first year of SKY. This came as a shock for SKY sellers, as they were expecting a supplementary income from SKY instead. They felt that the 'government had cheated them'. They were unable to comprehend that they were net consumers of power from MGVCL; i.e. they were consuming more power then what they produced. Eventually they did understand, but stopping or reducing the sale of water was not easily possible, due to their social compulsions. Hence, they kept the number of water buyers as well as their irrigated area constant, very few sellers actually paid up their electricity bills.

(c) *2nd year after SKY (2020–21)*

After two years of SKY, farmers were forced to make a gradual increase in water rates. The water rates increased from the earlier level of Rs. 40 per hour to Rs. 50–60 per hour. There was no change in the irrigated area or the number of farmers who were provided with irrigation service. The sellers were aware that their water buyers would neither be able to get water from elsewhere; nor get their own electricity connections easily. Water buyers were mostly one's relatives and neighbours, so 'how could they refuse to help them'? Besides, those farmers who were simultaneously buying and selling water at their different parcels had no option but to sell water on one parcel in order to be able to buy it on another. Hence, they continued to sell water as before, 'even while they were in a loss-making situation'. However, this 'loss' was only notional, as they still did not pay up their pending electricity bills, for which the pending amount increased cumulatively.

(d) *3rd yr after SKY (2021–22)*

In the third year onwards after SKY, slowly they realized that they were getting payable bills as they were net consumers of power instead of net producers; hence they would have to limit their water sale in order to curtail payable bills for electricity. They began to look for ways and means of curtailing their bills. They stopped selling water indiscriminately and started selling only to their immediate neighbours or relatives with whom they had long term relationships. They also increased water rates up to Rs. 120–140 per hour. They also realized that their solar power production was much lesser than their potential; hence they decided to put in more efforts in order to increase it. They gained awareness on the techniques of cleaning and upkeep of their solar panels, compared their daily power production with fellow farmers and learnt from each other. They also took help of their children and other youngsters in the village to gain more knowledge which could help them to increase their power generation. However, most water sellers still did not pay their electricity bills.

As shown in Table 7, most SKY sellers in the selected sample received payable bills under SKY. However, about 14% sellers who were earlier receiving payable bills had a turnaround and started receiving income from SKY.

Table 7: SKY Sellers' Experiences with SKY

Experience with SKY	*Proportion of SKY Sellers*
Earning income from power sale under SKY (Net Producers of Solar power)	07 (14.28)
Received payable bills under SKY (Net consumers of electricity)	34 (69.38)
Initially received payable bills (Net consumers); but later started receiving income from SKY (Net Consumers turned into Net producers)	07 (14.28)
Refused to comment	01 (2.04)
Number of Selected SKY Sellers	N=49 (100)

Source: Personal interviews by the author.

The respondents were shocked that they were receiving payable bills from instead of income from sale of solar power as expected. They reacted to this situation in a variety of ways. 91.83% respondents expressed that they had not paid up their outstanding bills, instead 'ready to go to jail', as the 'government has cheated us' by not giving us the 'subsidy that was promised to us'. 34.69% SKY sellers were either thinking of filing a case against MGVCL or had already complained to MGVCL about receiving payable bills. 38.77% of SKY sellers had increased their water rates, but retained water sale to the same number of buyers and were supplying the same number of irrigations to them as before. A smaller percentage of only 32.65% of the sellers had taken the route of maintaining the same water rates; and they were supplying the same number of irrigations as before, but had reduced the number of buyers that they were supplying water to. It was revealed that they were able to do so, as they were comparatively newer irrigation pumps who did not have age-old and long-established buyer-seller relations with their buyers. Hence, it was easier for them to stop selling water to some of them. A majority (79.59%) of the SKY sellers had resorted to increasing their water rates; while supplying to the same number of buyers, but reducing the number of irrigations from before.

SKY sellers expressed that they were given quite different information about the financial terms as well as benefits of SKY by the agencies at the time when they were being convinced to join the SKY scheme, which later turned out to be incorrect. Their expectations about incomes and gains from SKY did not materialize. Hence, they felt that the 'government has cheated them' and 'no one had benefitted from SKY as promised'.

Initially, when the payable bills were received from MGVCL, some farmers paid it up. However, after the first year of paying, when they found that 'other farmers had not paid their bills' and 'were expecting the government to waive them off in the future' (6.1%), they too decided to follow suit and did not pay their bills in the subsequent years. Few of them (8.1%) felt that 'if others will pay, then I will also pay'. They collectively represented their problems to MGVCL as well as local politicians many times, but felt that 'no one listened to the farmers'. Some sellers (10.20%) had stopped selling water altogether.

In the initial couple of years, irrigated land, number of buyers and water rates remained unchanged. From the 2ndyear onwards, gradually, water rates were increased, but irrigated land and number of buyers had remained unchanged. However, From the 3rd year onwards, not only were the water rates almost doubled, but the water charges also began to be collected from the buyers at the end of each season, instead of annually, as before. Also, the number of irrigations that were supplied, were also reduced to the extent possible. Now, water was not freely available to all buyers; but supplied only to close neighbours, relatives and those farmers who were in a desperate need. For instance, 'if the crop required 12 irrigations, only 10 irrigations would be supplied'. When asked whether this was not unfair to the buyers, they replied that 'it only affects the quality of the output marginally'. Moreover, 'it is better to reduce irrigations supplied to all existing buyers instead of stopping the sale of water to any 'one of them'.

Buyers from SKY sellers revealed that there had been almost 60% increase in water rates after SKY. Whereas before and immediately after the implementation of SKY, the price rise was nil to moderate, after receiving payable bills, the price increase was very sharp. Prices had increased in all three villages namely Ishnav, Ashapuri and Trambovad. However, buyers were happy that SKY provided electricity supply for 12 hours instead of 8 hours as before. Earlier, they had to book for water in advance and wait for their turn. After SKY, water was available more readily. Also, SKY brought electricity supply in the day-time which improved their convenience in irrigating their crops to a great extent. They 'didn't have to take the trouble of going to the farm at night'. Moreover, even though the buyers resented the increase in water rates, they informed that the area under irrigation as well as the number of buyers and quantity of water supplied to them had remained the same as before even under SKY.

Water buyers expressed a pertinent concern regarding increased water rates. They asked that as was well-known, SKY sellers did not pay their increased electricity bills. In that case, how is it justified to increase the water rates that they charged from their buyers? They expressed their helplessness by saying that as 70% of tube wells had increased water rates. Again, in the close-knit rural community, they could not shift easily from one seller to another, as this would deprive the seller from his source of income. Hence, they had to continue buying from the same seller even if for a higher water rate. Canal irrigation was not accessible in any of the study villages except Trambovad; where also, it was available only during monsoon. Hence, it was not possible for them to shift from a SKY seller to a non-SKY seller to combat the increase in water rates.

Table 8: Impact of SKY on Water Buyers

Advantages of SKY to Water Buyers	*Disadvantages of SKY to Water Buyers*
1. Water is available more freely because of increased hours of electricity supply 2. Freedom from inconvenience of night-time irrigation as was the case during JGY 3. More reliable electricity supply than what was during JGY	1. Water rates have increased immensely 2. If sellers are not paying their increased electricity bills, why are they charging higher water rates from us? 3. Option from buying water from Non-SKY seller was not easy; as 70% of connections had converted to SKY 4. Generations-old social ties with SKY sellers were difficult to snap over the increased price 5. Canal irrigation not accessible easily; either canal was too far away or canal water was available only during monsoon 6. Water charges were now collected at the end of every season; instead of the earlier practice of collecting them annually

Source: Data collected by the author during personal interviews.

Being rational humans, the water buyers would try to protect themselves from the increase in water rates and could either apply for their own electricity connection or use canal water for irrigation. Only 1 buyer out of the 36 who were interviewed, revealed that he had applied for his own electricity connection. Others had no plans to apply for a new electricity connection for various reasons. 44.4% revealed that they could not afford to do so due to lack of funds; while 30.5% stated that they owned so less amount of land, that it was not viable for them to construct a tube well. Few buyers (11.1%) also revealed that their land was owned jointly by many persons, all of whom did not agree on investing in the construction of a tube well.

Even though the study villages are included in the MRBC canal command area, in reality, canal water is not easily accessible to them due to various reasons. Firstly, canal water was inaccessible to most of the water buyers in Ishnav and Ashapuri and Rangaipura as they were at a great distance from the canal (83.3%). In Trambovad, the canal was relatively close by, but the terrain was such that flow irrigation through gravity was not possible (5.5%). Moreover, the canal was filled up substantially only during a good monsoon (83.3%). Hence, canal water was not a reliable source of irrigation, which meant that the water buyers had to primarily depend upon groundwater for their irrigation needs.

Non-SKY sellers did not increase their water rates post-SKY but kept them unchanged. This was possible because they still got electricity at a metered rate of Rs. 0.60 per unit as was the case under JGY. However, they felt that they were suffering a loss in income from the sale of water. As electricity supply was given to them only for 8 hours a day, they could sell lesser water compared to that by SKY sellers.

When asked about the reasons for opting for SKY, they expressed reasons such as:

1. They did not have clear records of land ownership in their own name. This was because often the land was inherited and there was more than one owner of the land. If one or more of the owners resided abroad, they would not be easily reachable for getting their signatures and completing all SKY-related formalities.
2. Sometimes, one of the joint-owners of the land was not convinced with the idea of joining SKY scheme.
3. It was also likely that in a jointly held land parcel, all the owners were not willing to contribute funds for the initial investment to be made under SKY; and hence, it could not be implemented.
4. The farmer was too poor to be able to raise funds for the initial investment under SKY
5. The farmer had meagre land due to which there was a lack of space for installing solar panels.

6. Most farmers owned more than one parcel of land. They may have an electricity connection and a borewell in one of the parcels, but had to buy water on other land parcels. Hence, they were sellers and buyers of water simultaneously on their various parcels of land.

Conclusion

Farmers expressed their dissatisfaction with all the regimes of electricity supply and price implemented till date, for various reasons. However, by far, their satisfaction with flat rate regime was highest in terms of price of power and with JGY, it was highest in terms of service provided. The water rates for irrigation service have been increasing periodically, but have risen very rapidly and almost doubled under the SKY regime. The SKY sellers were aggrieved because they were ill-informed about the financial arrangements of SKY and not properly oriented on how they could become net producers of power instead of net consumers. In course of time, they started learning and few farmers had also started earning net income from selling power under SKY. Majority of respondents did not pay up their electricity bills and had no idea how they would face the cumulative effects of their non-payment. However, they had shifted the burden on their water buyers either increasing the water rates, decreasing the irrigated land, decreasing the number of irrigations or stopping the sale of water altogether. Water buyers resented the fact that SKY sellers had increased the water rates in spite of not actually paying up the electricity bills owed by them. However, they were happy with SKY due to the increased hours of electricity supply and its more convenient timings. The non-SKY sellers did not increase their water rates after SKY. They felt left out because they could not sell as much water as the SKY sellers, since they had only 8 hours of electricity supply. Thus, while some aspects of SKY were welcomed, it has been the cause of a stark increase in water rates, which however, did not have a substantial impact on the area irrigated, number of water buyers or the amount of water supplied to them.

The author recommends i) capitalizing on the strengths of SKY i.e. increased hours of power supply under SKY; ii) make reforms in SKY in terms of rationalization of price of solar power offered to the power sellers so as to encourage them to become net power producers instead of new power consumers; and iii) make new SKY connections easy to acquire and affordable for those who are water buyers at present; particularly those with small amount of land. This could go a long way in expanding solarized irrigation, increase solar power production for the DISCOMS as well as make irrigation service more affordable for its sellers as well as buyers; which will ultimately contribute towards sustainable agricultural development in Gujarat.

References

[1] Bhatt, Sonal and S.S. Kalamkar (2017). "Solar Power Generation and Usage in irrigation: Lessons from a Novel Cooperative Initiative in India", Indian Journal of Economics and Development, Vol. 13, No. 3, Society for Economics and Development, Ludhiana, pp. 413–423.

[2] Closas, Alvar and Edwin Rap (2017). 'Solar-based groundwater pumping for irrigation: Sustainability, policies and limitations', Energy Policy, Volume 104, May 2017, Pages 33–37,accessed from Solar-based groundwater pumping for irrigation: Sustainability, policies and limitations—Science Direct, https://doi.org/10.1016/j.enpol.2017.01.035

[3] Dubash, Navroz K. (2000). 'Ecologically and Socially Embedded Exchange: 'Gujarat Model' of Water Markets, Economic and Political Weekly, Vol. 35, No. 16 (Apr. 15–21), pp. 1376–1385.

[4] Dubash, N.K. (2002). 'Tube well capitalism: groundwater development and agrarian change in Gujarat', pages 300, Oxford University Press, New Delhi, India ISBN: 0195657470.

[5] Government of Gujarat (2022). Official website of the Narmada, Water Resources, Water Supply and Kalpasar Department, Government of Gujarat, accessed from https://guj-nwrws.gujarat.gov.in/showpage.aspx?contentid=2953&lang=Gujarati on November, 01, 2022

[6] Government of India (2007). 'Report of the Expert Group on Groundwater Management and Ownership', Planning Commission, Government of India, New Delhi, September, 2007, pp: 32–33.

[7] Kolvalli, S. and Chicoine, D.L. (1989). 'Groundwater markets in Gujarat, India', International Journal of Water Resources Development, Vol. 5, Issue (1), pp: 38–44. https://doi.org/10.1080/07900628908722410

[8] Kumar, M. Dinesh and O.P. Singh (2001). 'Market instruments for demand management in the face of scarcity and overuse of water in Gujarat, Western India', Water Policy, Volume 3, Issue 5, pp: 387–403, https://doi.org/10.1016/S1366-7017(01)00081-2

[9] Lefore, Nicole, AlvarClosas and Petra Schimitter (2021). 'Solar for all: A Framework to deliver inclusive and environmentally sustainable solar irrigation for small-holder agriculture', Energy Policy, Vol. 154, July 2021, https://doi.org/10.1016/j.enpol.2021.112313

[10] Manjunatha, A., Speelman, S., Chandrakanth, M. and Van Huylenbroeck, G. (2011). Impact of groundwater markets in India on water use efficiency: A data envelopment analysis approach. Journal of Environmental Management, 92(11), 2924-2929. https://doi.org/10.1016/j.jenvman.2011.07.001

[11] Shah, Tushaar and VishwaBallabh (1997), 'Water Markets in North Bihar: Six Village Studies in Muzaffarpur District', Economic and Political Weekly, Vol. 32, No. 52 (Dec. 27, 1997 – Jan. 2, 1998), pp. A183-A190.

[12] Shah, T., Bhatt, S., Shah, R. and Talati, J. (2008), 'Groundwater governance through electricity supply management: Assessing an innovative intervention in Gujarat, western India, Agricultural Water Management, Volume 95, Issue (11), pp. 1233–1242.

[13] Shah, Tushaar and S.D. Chowdhury (2017), 'Farm power policies and groundwater markets: Contrasting Gujarat with West Bengal (1990–2015)', June 2017, Economic and Political Weekly Vol. 52, Issue (25), pp. 39–47.

[14] Shah, Tushaar, Gyan Rai (2021). 'Solar Pumps and Water-Energy Nexus in Gujarat, India: First Assessment of the World's Largest Pilot on Grid-connected Solar Irrigation Pumps'; preprint; not been peer reviewed by a journal, https://doi.org/10.21203/rs.3.rs-658617/v1, This work is licensed under a CC BY 4.0 License; accessed from https://www.researchsquare.com/article/rs-658617/v1, posted on 14 Jul, 2021.

[15] Shah, Tushaar, Abhishek Rajan, GyanPrakash Rai, Shilp Verma and NehaDurga (2018). 'Solar pumps and South Asia's energy-groundwater nexus: exploring implications and reimagining its future', published 1 November 2018, IOP Publishing Ltd, Environmental Research Letters, Volume 13, Number 11, Focus on Interactions Between Science and Policy in Groundwater Systems.

[16] Singh, Dalbir (2007). 'Who Gains and Who Loses in the Game of Groundwater Markets in Water-scarce Regions', Agricultural Economics Research Review, Vol. 20, July-December, 2007, pp: 345–360.

[17] Tamuli, Jitu, Mrinal Kanti Dutta and Payel Priya Kashyap (2021). 'Groundwater Markets in Water Abundant Conditions: Factors Affecting Farmers' Decision to Sell Water in Assam in North East India, Water Economics and Policy, Volume 07, No. 04, 2150015 (2021), https://doi.org/10.1142/S2382624X21500156

Water Security for Agriculture Sector Reuse & Recycling of Treated Sewage for Agriculture

Atul G. Shroff[1] and Mandar Prabhune[2]

Preface

1. Water scarcity in India is expected to worsen as the overall population is expected to increase to 1.6 billion by year 2050.
2. *Pradhan Mantri Krishi Sinchayee Yojana* (PMKSY) was launched during the year 2015–16, with an aim to enhance physical access of water on farm and for expanding cultivable area under assured irrigation, improve on-farm water use efficiency, introducing sustainable water conservation practices, etc.
3. Consuming over 80% of the country's freshwater, agriculture in India is one of the key areas being disrupted by climate change. Farmers often rely on groundwater resources to secure their crops against uncertain weather patterns.
4. According to the Water Resources Group (WRG), by 2030, India will only have 50% of the water it needs. This imminent crisis is likely to be the worst in India's history and will disproportionately impact agriculture.
5. The global food production therefore needs to make better use of water resources and avoid wastage; produce more food per litre of water is the key
6. Water reuse and recycling is part of the solution and a crucial factor to implement a sustainable management of the water cycle.

Reuse & Recycling of Treated Sewage

The greatest challenge in the Water and Sanitation sector over the next two decades will be the implementation of 'Low-Cost Sewage Treatment' that will at the same time permit selective REUSE of treated effluents (disinfected & detoxified) for agricultural & industrial purposes.

---

[1] Transpek Industries Ltd.

[2] Technical Head, Transpek Industries Ltd.

The Simple Solution

- Treatment of waste water using a simple, decentralized, & environment friendly methods using minimal energy which is applicable to rural as well as urban areas.
- Avoid a long transmission of wastewater to the centralized sewage treatment systems, thus saving energy, high maintenance cost and making available the treated water at decentralized places.
- Reuse of treated water for various purposes especially in agriculture & industry—a sustainable source of water.

Shroffs Foundation Trust in cooperation with UNICEF, 42 villages of Gujarat were surveyed. Based on the survey, it was found that around 0.5 MLD of waste water is available for reuse purpose which is otherwise disposed of in nearby nalla.

Shroff Foundation Trust									
Template for village level data collection for preparation of sewage Treatment Plants									
1	*2*	*3*	*4*	*5*	*6*	*7*	*8*	*9*	*10*
Sr. No.	*Name of the Village/Gram Panchayat*	*Population*	*Total Household*	*Total water supplied in the village*	*Percentage Coverage Of Drainage Network*	*Nature of Wastewater*	*Waste Water Generation in the Village*	*Area*	*Recommended Plant Capacity*
		Persons	*Number*	*Liter/Day*	*%*		*Liter*	*Sq.m.*	*KLD*
1	Nandol	6966	2125	460000	95	Grey water	236992	47800	240
2	Aadarajmoti	13273	2660	800000	90	Grey water	412160	NA	420
3	Unava	8842	2050	700000	95	Grey water	360640	NA	360
4	Rupal	8674	2145	470000	95	Grey water	242144	NA	260
5	Nardipur	9813	3100	650000	90	Grey water	334480	1000	340
6	Balva	8304	1400	500000	95	Grey water	257600	2500	260
7	Saij	21505	5959	840000	90	Grey water	432768	520	440
8	Lodra	10126	2400	800000	90	Grey water	412160	NA	420
9	Ajol	8280	2317	370000	90	Grey water	190624	2400	200
10	Itadara	14375	3411	650000	90	Grey water	334880	9600	350
11	Salal	10546	2537	400000	90	Grey water	206080	485	220
12	Punsari	7475	1637	375000	90	Grey water	193200	1818	200
13	Kathi Devaliya	1725	280	90000	90	Only Grey water	46368	625	50
14	Bhesan	15755	2800	1000000	90	Only Grey water	515200	NA	515

Shroff Foundation Trust									
Template for village level data collection for preparation of sewage Treatment Plants									
1	*2*	*3*	*4*	*5*	*6*	*7*	*8*	*9*	*10*
Sr. No.	*Name of the Village/Gram Panchayat*	*Population*	*Total Household*	*Total water supplied in the village*	*Percentage Coverage Of Drainage Network*	*Nature of Wastewater*	*Waste Water Generation in the Village*	*Area*	*Recommended Plant Capacity*
		Persons	*Number*	*Liter/Day*	*%*		*Liter*	*Sq.m.*	*KLD*
15	Chodvadi	7245	1400	350000	90	Only Grey water	180320	NA	180
16	Mendarda	21735	4020	1500000	90	Both black & grey water	1380000	NA	1380
17	Jam Kandorana	18745	3200	1000000	95	Both black & grey water	920000	1140	920
18	Virpur	13685	2433	900000	90	Both black & grey water	644000	More 300	645
19	Saapar	13800	2958	700000	90	Only Grey water	360640	250	360
20	Pardi	10810	2060	500000	90	Both black & grey water	460000	800	460
21	Paddhri	13279	3760	1000000	95	Both black & grey water	920000	1500	920
22	Kunjrao	9200	1800	550000	95	Grey water	283360	NA	285
23	Adas	16905	4000	950000	92	Grey water	489400	NA	490
24	Vasna (Borsad)	7073	1050	380000	100	Grey water	195776	2322	195
	Al-Madina nagar	3565	450	195000	85	Grey water	100464	NA	100
25	Virsad	11500	2083	687000	92	Grey water	353942	2500	355
26	Dahewan	15468	3750	850000	90	Grey water	437920	NA	440
27	Ralaj	11011	1750	660000	96	Grey water	340032	60000	340
28	Undel	17825	3200	1000000	95	Grey water	515200	10000	515
29	Nar	9200	2083	520000	96	Grey water	267904	1393	268
30	Dharmaj	14116	2627	840000	97	Both black & grey water	772800	142	772
31	Dabhou	7820	1400	400000	94	Grey water	206080	8094	206
32	Thamna	7015	1300	360000	96	Grey water	185472	800	185
33	Valod	23545	2900	800000	50	Both black & grey	736000	-	750

Shroff Foundation Trust									
Template for village level data collection for preparation of sewage Treatment Plants									
1	*2*	*3*	*4*	*5*	*6*	*7*	*8*	*9*	*10*
Sr. No.	*Name of the Village/Gram Panchayat*	*Population*	*Total Household*	*Total water supplied in the village*	*Percentage Coverage Of Drainage Network*	*Nature of Wastewater*	*Waste Water Generation in the Village*	*Area*	*Recommended Plant Capacity*
		Persons	*Number*	*Liter/Day*	*%*		*Liter*	*Sq.m.*	*KLD*
						water			
34	Vagra	10658	1871	650000	85	Grey water	334880	3718	350
35	Mota	11500	2561	400000	100	Both black & grey water	368000	12000	400
1	*2*	*3*	*4*	*5*	*6*	*7*	*8*	*9*	*10*
36	Kadoda	18223	4725	900000	95	Grey water	463680	9000	500
37	Hajira	22425	3500	750000	95	Grey water	386400	70000	400
38	Kholpad	28750	5200	1000000	90	Both black & grey water	920000	2000	950
39	Laskana	34500	5841	900000	95	Both black & grey water	828000	2000	850
40	Mota Borsara	8050	1635	500000	95	Grey water	257600	4700	300
41	Pipodara	10972	1110	400000	94	Both black & grey water	368000	9500	400
42	Kim	32200	7989	1300000	95	Both black & grey water	1196000	150	12 MLD
		576479	117477	29047000	3942		19047166	268757	18191

Biofilter Technology offers a unique solution, which is environment friendly, cost-effective and provides multiple benefits.

About Biofilter Technology (Patent No. 393264)

- The concept is *"Design with Nature"* how to work cooperatively with nature, while at the same time learning the tools of modern science.
- *Bio-filter Technology* is a synthesis process, which harnesses the energy, carbon and other elements of the waste and converts them to precious *"Bio-nutritional"* products like energy rich humus & bio-fertilizer and nutrient rich water.

- The process utilizes the ability of the earthworms and beneficial microbes *(transzyme)* to break down organic waste present in the wastewater and transform it into worm-cast.
- It involves removal of organic matter by adsorption and filtration followed by biological degradation and oxygen supply by natural aeration to the treatment system.

Green Technology transzyme
to sustain Environment

Comparative

Conventional STP/ETP	*Vermifiltration Technology*
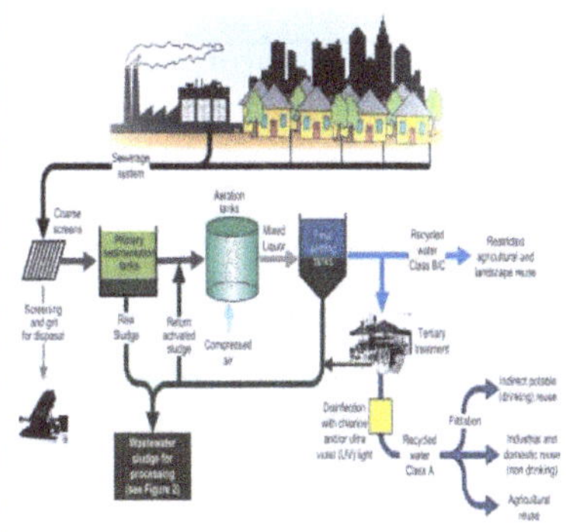	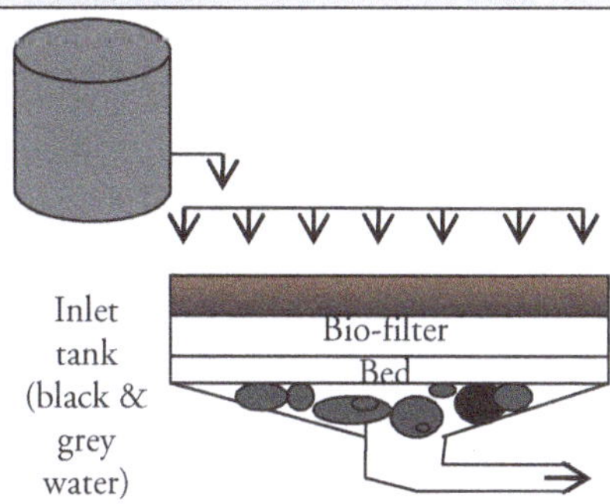

Advantages of Trans-Biofilter Technology

1. Low Operating and Maintenance Costs;
2. Low ENERGY input with Low Emission of Greenhouse Gases;
3. No SLUDGE is formed which are HAZARDOUS wastes requiring safe disposal at high cost. All solids in sewage are eaten by earthworms & excreted as VERMICOMPOST;
4. Aerobic & HYGIENIC process & hence no ODOUR problems;
5. Substantial reduction in colour;
6. Able to handle high COD & BOD loads;
7. Able to meet POLLUTION norms for safe discharge in RIVERS & Lakes;
8. Treated water is NUTRIENT RICH and useful for increasing natural SOIL FERTILITY for high agricultural production and significantly reducing artificial fertilizers.

Case Study of Excel Cropcare Ltd., Bhavnagar, Gujarat, India

Snapshot

Customer: Excel Cropcare Ltd., Bhavnagar, Gujarat.

Challenge: To Fulfil the Requirements of Water for Agriculture & Industry in the Situation of Growing Scarcity of Freshwater Resources all over World. Global Agriculture use 85% of total Freshwater (Ground & Surface Waters) on Earth which is a Depleting Resource.

Solution Offered: The Transchem Agritech Pvt. Ltd., will Use its Innovative "Bio-filter (Vermifilter) Technology by Earthworms" to treat Municipal Sewage and Re-use the Clean & Purified (Disinfected, Detoxified & also Nutritive) Water for Agriculture & Industries.

Economic & Environmental Benefits:
1. Biofilter is 'Low-cost Green Technology' for Sewage Treatment.
2. Reuse of Community Wastewater for Industrial & Agriculture purposes.
3. Huge Conservation of Freshwater Resources which is a Depleting Resource on Earth.
4. Considerable Saving in Cost of Operation & Maintenance.
5. High Quality Water at Affordable Price.
6. Huge Earthworms Biomass & Vermicompost results as By-products to be used by Farmers in Farms for High Food Production Eliminating the Dangerous Agrochemicals proving to 'Slow Poison' for Farmers & degrading the Farm Soil.

We have successfully installed 800 KLD (Recently expanded to 1200 KLD) Bio-filter (Vermifilter) Plant at ECCL, Bhavnagar, Gujarat in 2011 for treatment of City Sewage Water and the reuse of treated CLEAN water in Industries & Agriculture saving huge groundwater which is a depleting resource.

The basic idea for installing "Trans-Bio-filter" plant was to use City Sewage as a source of usable Water by converting them into 'disinfected & detoxified' high quality WATER. Since Bhavnagar is water scarce area and also being nearer to Sea Coast, the quality of groundwater is of high TDS (700–800 ppm), which makes it non-suitable for direct use in process systems. The cost of potable water in Bhavnagar is to the tune of Rs. 50–60 per KL.

Average Results Obtained (Bhavnagar)

Parameters	*Before Treatment*	*After Vermifilter Treatment*	*After R.O. Treatment*	*Standard Norms for Land Irrigation*
pH	7.0–7.8	7.5–8.0	7.0 – 7.5	5.5–9.0
TDS (ppm)	1500–2000	1200–1800	< 200	2100
Turbidity (NTU)	40–100	<10.0	Nil	
COD (ppm)	200–400	50–80	< 20	100
BOD (ppm)	100–200	10–20	< 5	30
DO (ppm)	Nil	3–4		
Colour	Dark Grey	Pale Yellow	Clear White	
Odour	Strong	Odourless	Odourless	
Bio-Assay	–	100% survival of fish after 96 hours in 100% effluent		90% Survival of Fish After 96 Hrs. in 100% Effluent
Faecal Coliforms (MPN/100 ml)	$> 10^6$	$< 10^3$	< 100	$< 10^3$

trans BIO-FILTER WASTE WATER TREATMENT TECHNOLOGY

TRANSCHEM AGRITECH Pvt. Ltd.

Comparison of Influent and Effluent from Bhavnagar STP

BEFORE TREATMENT AFTER TREATMENT

21

Reuse of Treated Water for Agriculture

"Resource of Nutrients"

Criteria for Sustainability in the Treatment of Wastewater for REUSE

1. No use of CHEMICALS in wastewater treatment & cleaning;
2. No DILUTION of high strength wastewaters with CLEAN WATER.
3. Maximum RECOVERY & RE-USE of treated clean water in agriculture which uses 85% of earth's water globally;
4. Application of EFFICIENT, ROBUST & RELIABLE treatment technologies, which are 'LOW COST' in construction, operation and maintenance leading to a high 'self-sufficiency' in all respects;
5. Acceptable to the local people & residents.

Nutrient Conversion in Vermifiltered Sewage Water Making it More Nutritive for Crop Irrigation & High Farm Productivity

Parameters	Before Treatment	After Treatment
NH4 -N(ppm)	25-40	<1.0
NO3 (ppm)	10-20	>50
Total P (ppm)	4-8	1-2
Available P (ppm)	1-2	5-7
Potassium (ppm)	10-15	20-25

Considering the nutrient content in treated water as:
- Available Nitrogen - 50 mg/l
- Available Phosphorous -- 7 mg/l &
- Available Potassium - 25 mg / l

And assuming 800 KLD treated water availability, the nutrient availability would be:
Nitrogen - 40 Kgs/day i.e. 14000 Kgs/annum eq. to 28 Tons of Urea
Phosphorous - 5.6 Kgs/day i.e. 2000 Kgs/annum eq. to 4.0 Tons of DAP
Potassium - 20 Kgs/day i.e. 7000 Kgs/annum eq. to 12.0 Tons of Potash

As discussed above, the Vermifiltered Sewage is also highly 'NUTRITIVE', high in bio-available 'NITROGEN', 'PHOSPHORUS' and 'POTASSIUM' good for crop growth & productivity. The best part is that the FOOD Crops produced are completely ORGANIC.

Decentralize Waste Water Treatment Plants for Villages

Case Study – KVK Jalna, Aurangabad, Maharashtra, India

- Jalna is a city located in Marathwada and receives very less rainfall and is drought prone area. Having population of around 3.0 lakhs, availability of domestic sewage is sufficient including in summers.
- KVK Jalna was facing huge shortage of water for agriculture and therefore, it was decided to treat the city sewage (from nearby nalla) through Biofilter Technology and reuse for agriculture.
- Since then the KVK is able to take 3–4 crops per year and their problem of water shortage was completely resolved.

Decentralize Waste Water Treatment Plants for Villages

Case Study – Khanderaopura Village, Vadodara, Gujarat India

- Khanderaopura village (Padra block of Vadodara District, Gujarat) is a village having population of 1500 having 185 households.
- In 2016 this village had an outbreak of Cholera and it earned a bad name. It remained focus of media for a long period. Shroff Foundation Trust was requested to look into the problems of sanitation and hygiene of this village.

- Thus, it was decided by the village community that this village would need to be qualified for ODF and the drainage line would be repaired to make it functional.

Reuse of Treated Sewage

- Khaderaopura Village generates 100 to 120 KLD domestic waste water @ 70 L/person/day.
- Proper mapping was done to collect the entire quantity of waste water generated for further treatment.
- A solar based biofilter plant of 120 KLD capacity was installed. The total treated water is used by villagers for agriculture.
- The plant is operated by the villagers after providing training. This has enabled them to take 3–4 crops per year.

Promoting Water Security for Resilient Agriculture We Need the Blue to Keep the Green: Development Support Centre Marching Towards Its 29 Years Journey in Promoting Water Security for Resilient Agriculture

Mohan Sharma[1]

Development Support Centre (www.dscindia.org) is one of the pioneer Civil Society Organizations of India that is dedicated to promoting Participatory Natural Resource Management and Sustainable Rural Livelihoods. The organization was founded in 1994 by the Late Shri Anil C Shah (retired IAS, Gujarat) as per the demands of various stakeholders and eminent individuals concerned with the state of rural development in Gujarat. DSC works directly in both rainfed and irrigated agriculture areas to facilitate the rural communities especially the vulnerable and deprived in solving their collective problems by themselves in a sustainable manner.

DSC`s experience in community-led supply and demand side management of water in agriculture for promoting water security.

28 years Journey towards Water Security

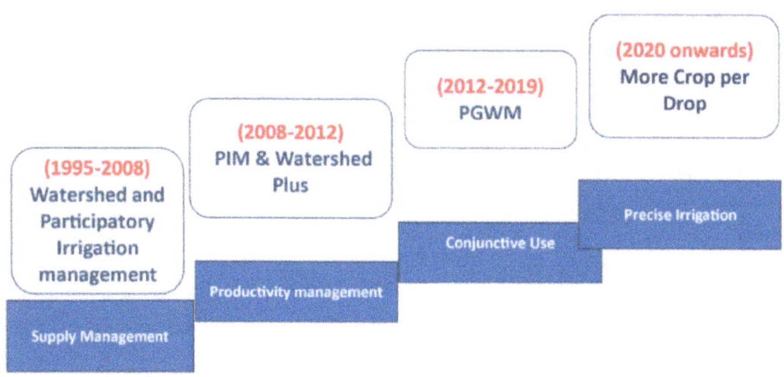

[1] Executive Director, Development Support Centre.

Water has been the centre of all DSC`s developmental initiatives since the beginning. Responding to the local needs, DSC joined hands with the government departments in piloting 10-village watershed project in Dhari taluka, Amreli district and Participatory Canal Irrigation Management project in Thalota village of Visnagar taluka in Mehsana district Gujarat to improve access to water for irrigation and domestic uses. These pilots proved very useful to the community to access water with better equity, efficiency and financial viability. The watershed and PIM initiatives were upscaled in collaboration with the Water Resource Department and Rural Development Department of Gujarat through policy and practice in diverse geographies.

2008 to 2012

DSC and its partners realized a need to enhance agriculture productivity by launching Watershed Plus and PIM Plus initiatives to stabilize and increase farm incomes through the extension of better crop management practices and low-cost technologies. These initiatives succeeded in reducing the cost of cultivation, increasing crop yields, mitigating man-made risks and better price realization by marginal and other farmers. However, in the race for production growth and income diversification, a large number of CBOs in Gujarat started reporting about over-exploitation of water resources, especially deep groundwater (up to 700 feet in North Gujarat) in both rainfed and canal command areas to DSC which may cause serious threat to the viability of agriculture-based livelihoods in these regions.

2012 to 2019

To check the over-exploitation of groundwater in rainfed areas, 2012 onwards DSC launched two pilot projects including MARVI (Management of Aquifer through Village level Recharge Initiatives in 6 watershed villages in Meghraj block of Aravali district Gujarat in collaboration with Western Sydney University, Australia. This project helped in trials of community-led monitoring and management of groundwater and to sensitize the tribal and community that the groundwater is not unlimited.

It also launched an Integrated Water Resource Management (IWRM) through the conjunctive use of canal and groundwater in 3 water scarce govt irrigation schemes of Dhari, Guhai and Mazum in North Gujarat.

These initiatives on PGWM and PWRM were successful in promoting inclusive community-based organizations that prepared and implemented village-level water budgets and water security plans. The village-level *Sujal Samitis* leveraged govt. schemes for rainwater conservation, groundwater recharge and efficient irrigation technologies and succeeded in filling the gap between supply and demand of irrigation water in 24 pilot villages. The villagers learned to safeguard deep aquifers from further exploitation by supply and demand side management of surface and shallow groundwater resources in a conjunctive manner. The CSR partners, donors and Government departments support these initiatives.

2020 Onwards

The final layering to DSC efforts for promoting water security has been done through a pilot Innovation Project namely *More Crop Per Drop* under Govt. *Atal Bhujal Yojana* in Dark zone areas of Mehsana district, Gujarat. This project is designed based on the successful results received from the Sugarcane Intensification Initiative (SII) and Onion Intensification (OII) Initiatives carried out by DSC, its CSR partner and Sugarcane Cooperatives in the Kukdi Irrigation Complex Ghod River basin project of Maharashtra. The sugarcane intensification initiative (SII) demonstrated successfully that the sugarcane farmer can save more than 5000 cubic meters of water by adopting low external inputs and sustainable soil-irrigation-crop input management practices without any additional cost to conventional farming. More than 10,000 farmers have switched to SSI and OII and saved millions of cubic meters of water.

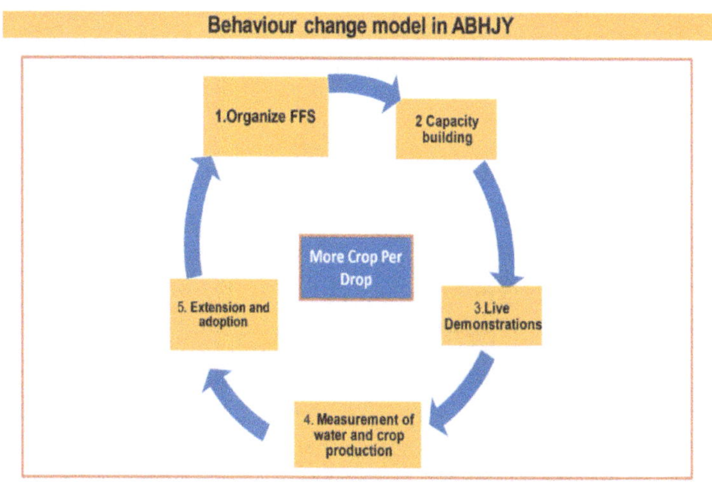

The MCPD aims to promote mass behaviour change of farmers towards on-farm water management through a simple and low-cost model: Design-Demonstrate-Monitor-Measure-Analyses-Interpretation-Extension-Adoption of 'Low Input High Output' package of practices in local water guzzler crops like cotton, Castor, wheat, summer millet and summer sorghum etc.

Experience of More Crop Per Drop project in Mehsana district Under Atal Bhujal Yojana

The innovative project, More Crop Per Drop (MCPD) has been launched in the five most critical villages of Davada, Butapaldi, Chhathiyarda, Devinapura and Deloli in Mehsana block, Gujarat. Farmer Field Schools were formed and capacitated through exposure visits and training. Live demonstrations of More Crop Per Drop (combined efficient irrigation technologies and natural farming practices) were designed in collaboration with *Krushi Vigyan Kendra*, Agriculture Technology Management Agency (ATMA), Spice Research Station, Jagudan, Wheat Research Station, Vijapur, etc.

Water measuring devices like flow meters and cut-throat flumes were installed on farmers' fields to measure the accurate flow of piped and gravity flow water. Crop water requirement is assessed regularly through moisture meters with sensor technology and rainfall gauges. The soil was tested, demo was given about soil mulching and water measurement for better crop management.

The baseline of soil health and water quality indicated high salinity and therefore, the farmers were motivated to replace chemical fertilizers with biofertilizers and biopesticides to improve overall soil carbon and humus content. High-temperature tolerant and less water-intensive seed varieties were selected and given to farmers under the package of practices. The farmers saw positive results as 22% less water was used in the MCPD demo plot of summer fodder (Pearl Millet) compared to the control plot. Moreover, the yield produced was 18% more on the demo plot than on the control plot. Similar results were found in summer fodder (Sorghum) with a 36% saving of water and a 22% increase in production.

The pilot project when ends in 2025, will give an insight into the benefit of using More Crop Per Drop plan but the need of the hour is to implement policies at the district level, State and Central levels along with the Panchayat Institutions, farmers and animal rearers to overcome the dearth in supply versus demand of water to achieve climate resilience and sustainable livelihood.

Key achievements

DSC promoted sizeable village institutions to upscale its learning on Integrated Water Resource Management towards achieving village-level water security programs in four western states of India. Today, DSC reaches out to more than 1000 villages in 31 blocks of 15 districts of Gujarat, Madhya Pradesh, Rajasthan and Maharashtra through direct field initiatives in collaboration with government departments, CSR and other partners.

Integrated Watershed Management

DSC has facilitated the rainfed area communities in treating 75,000 hectares of degraded watershed area cumulatively for the revival of landscape and vegetation cover along with the creation of 1,095 structures creating 59.21 lakh cubic meter storage potential and over 3,000 ha supplementary irrigation potential, leading to increased access to water and 10–22% growth in crop yields. The organization has formed more than 130 village watershed committees in four states in partnership with government departments and CSR partners. The village communities have also restored 12,234 ha of critical habitat in 112 villages for endangered flora and fauna through the regeneration and protection of indigenous plants.

Participatory Groundwater Management

Under *Atal Bhujal Yojana* in Mehsana district, Gujarat 132 Participatory Groundwater Management committees, 59 village *Sujal Samitis* and 1 registered groundwater cooperative have been promoted and facilitated in the water-scarce areas of four states. They are involved in water budgeting and implementing water security plans through convergence and sharing groundwater among themselves without drilling new borewells.

Participatory Canal Irrigation Management

DSC has developed physical and social capital under PIM and watershed by Canal Rehabilitation, Water User Association. It has facilitated more than 400 WUAs, mainstreaming the tail-end, marginal and women farmers in planning and managing the operations of canal irrigation and more than 200 WUAs have become financially viable.

The results before PIM show, that off-farm (conveyance and distribution system) irrigation efficiency is much higher which was around 0.34 million cubic feet per hectare during the pre-PIM phase of 1980–1994 which lowered to 0. 23 million cubic feet per hectare during the post-PIM phase of 1995–2010. Hundreds of farmers received water first time in history after WUA took over irrigation management of minor canals in their hands.

Conclusion and Way Forward

In India, about 80% of current water is used in agriculture. The irrigated area accounts for nearly 48.8% of 140 million hectares of agricultural land in India. The remaining 51.2% is rainfed. There is a considerable gap between the potential created and the actual utilization of irrigation.

In the rural areas of Gujarat, there has been tremendous growth in agriculture, dairy and irrigation infrastructure in the last decade (2008–2018). However, the growth will not be able to reach its full potential due to the increasing gap between supply and demand for water.

There is an urgent need to re-initiate participatory approaches and integrated interventions in meeting demand with supply in terms of both quantity and quality of water. This is possible to achieve through the launching of appropriate missions to change the water use behaviour of the farmers, incentivizing demand side management of water in irrigation and promoting conjunctive use through integrated planning, implementtation and monitoring of various govt. programmes and policies such as ABHJY, NJJM, PMKSY, ATMA, AIBP and MNREGS, etc.

Recommendations for Farmers and Animal Holders

1. Form farmer collectives for enhancing agriculture production, adopting extension technologies and collective marketing in a three-tier structure.
2. Adopting integrated farming systems and climate change resilient varieties and breeds for securing resilience.
3. Adopt water-sharing measures through conjunctive use of surface and groundwater and stop drilling deep tube wells.
4. Enhance soil health by producing and using bio inputs at home.
5. Make use of appropriate irrigation practices. This may not necessarily be drip, other methods like furrow irrigation are equally effective.
6. Improve access to free apps for crop, weather and market advisory.
7. Do responsible farming and gradually shift towards natural farming.

Implementation by District-Level Departments

1. From the District Irrigation Management Agency for anchoring demand and supply side management of water for achieving water security in the district and to help the Panchayats carry out water security plans and conduct water audits.
2. Make GPs accountable for maintaining water Inventory charter by village counsels mandatory and provide positive incentives for Panchayats for the same.
3. Launch Micro Irrigation Mission and involve NGOs in water literacy and build local community resource persons cadres,
4. Undertake the repair of water conservation and harvesting structures through MGNREGA and Village Action Plan.
5. Promotion of Bio Input Production and Resource Centres under *Kisan Samman Nidhi* by GoI to switch towards natural farming methods gradually.
6. Promote roof water harvesting under an innovative scheme like *Ladli Bahan* of MP where there is a problem of fluoride contamination in groundwater and reduce the same through recharge of the shallow aquifer.
7. Giving mandate and target to WALMI and engaging competent NGOs and another training institute to develop a cadre of water worriers/ water champions/ *Jal Jankars* under the scheme of *Jal Shakti Mantralay* and promoting village-level inclusive water institutions.
8. Enhance the participation of women in decision-making, water budgeting, water security plans and water governance at the village level.

Panchayat Institutions

1. Strengthen and incentivize the village institutions such as *Paani Samities* and WUAs and pilot a more holistic *Sankalit Jal Vyavasthapan Samities* (IWRM) that works on all water resources and all water uses in the village.
2. Carry out regular repair, maintenance and management of water harvesting structures and soil and moisture conservation measures undertaken through various schemes.
3. Execute recharging of wells and tube wells under the guidance of District administration.
4. Implement measures for greywater treatment and recycling.

Recommendation for Government

1. Realign the programs and policies to create an enabling environment for facilitating water users, practitioners and donors in achieving water security.
2. Launch a water Security Mission in the state to promote decentralized water governance in line with *Atal Bhujal Yojana*. The water security plans approved by GP must be considered as the perspective plan to spend govt. grants on water resource development and agriculture livelihood.
3. Promote the Integrated Water Resource Management (IWRM) Model in all the canal commands in the state for the designed promotion of conjunctive use of surface-canal-groundwater.
4. Align the PM *Kisan Samman Nidhi* to promote low water-guzzling crops such as millets, pulses and spices
5. Launch of Watershed 3.0 projects especially dedicated to large-scale maintenance and repair of structures and artificial recharge.
6. Reduce water consumption in Dairy by promoting precise irrigation, low water requirement fodder crops such as millets, three-tier farming in tribal and other areas.
7. Promote roof water harvesting under innovative schemes like *Ladli Laxmi* in MP particularly in hotspot areas of water contamination at different scales.
8. Revisiting the performance of canal command areas and their institutional structure and making them responsive to the present needs of collaborations.

Setup Moisture Meter, Flow Meter and Rain Gauge

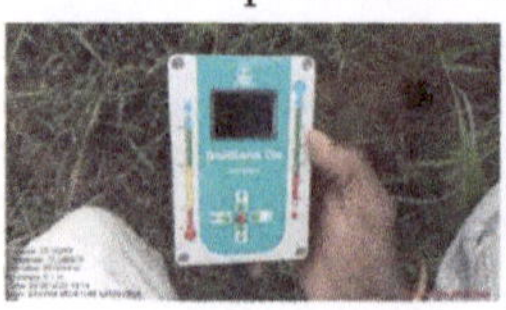

Soil Moisture Meter with sensor

Flow meter Reading

Measurement of Daily rainfall by Field Supervisor

Setup Irrigation Water Measurements

Flow meter Installation

Cut throat Flume Installation

Water used in Irrigation measured by Cut throat Flume

Soil Health Improvement

Bio Pesticides

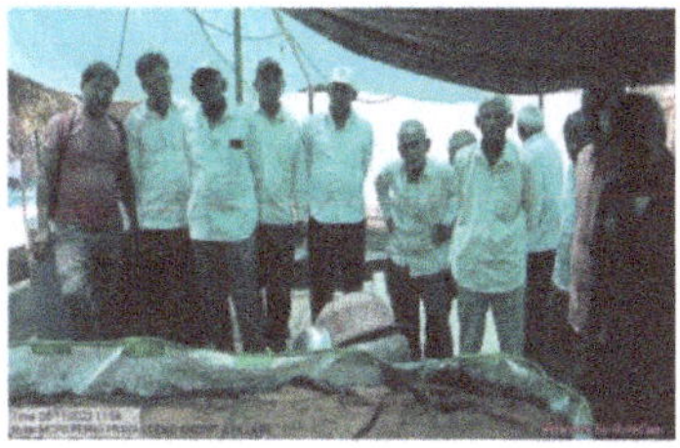

Bio fertilisers

Some Glimpses from Initial Stage of the Project

No of plant calculation was being done in 1m * 1m area

Vegetative growth Measurement by Field Facilitator

Vermi Compost Distribution to farmer

SEWA's Natural Resource Management Campaign: Water, Women and Work

Bhartiben Bhavsar[1]

S*washrayi Mahila Sewa Sangh*" SEWA is an organization of poor women who is unorganized and working together in 14 districts of Gujarat. Among the total membership of SEWA 54% (3,70,627) members are engaged in animal husbandry and agriculture. Be it agriculture or animal husbandry or daily life...water is an important factor of life. Life is impossible without it.

The world is far from water security. In many parts of the world the demand for water is already much greater than the available supply. Water for agriculture and, therefore, food is not given sufficient attention at the global stage, where water supply and sanitation issues currently dominate.

The districts most affected by water insecurity include Rajkot, Bhavnagar, Amreli and Jamnagar in the Saurashtra region; and Banaskantha, Surendranagar, Patan and Mehsana in north Gujarat. In several villages of Patan and Banskanatha, the ground water level has dropped and contents of fluoride in the water has increased alarmingly, posing health hazards to the community.

In agriculture, crop needs frequent watering, if we talk about the water requirement of the plant, then plant just need a moisture for grow, but for years we have been using the furrow irrigation system and due to which water is wasted a lot and not all the water can be used by the plant. Gradually the water table is also getting lower and if no specific measures are taken for it, the shortage of potable water is imminent.

There are also many technologies for irrigation and sustainable cultivation practices so that water can be saved but these methods do not reach the farmer and the cost of the technology is too exorbitant to adopt by the farmer.

---

[1] Rural Development Department, Clena Sky Campaign, Self Employed Women's Association (SEWA),

If there is no water, the seriousness of its impact on human life is not beyond imagination. Therefore, awareness about its significance need to be explained to the people.

SEWA works with so many small and marginal farmers. Small and marginal farmers have very little land. Therefore, they do not have the facility of their own borewell for irrigation and often the crops fail due to lack of adequate water resources. Technology is not easily reach out at the small farmers. To cope with these problems, the SEWA started water campaign in 1995. The problem for drinking water and agricultural irrigation water especially in the remote part of the desert area of Patan district needs no underscoring. Therefore, SEWA formed a water protection committee in village that discussed the water related issues along with their solutions.

The main issues of the village level were as follows:

- Scarcity of irrigation water for agriculture as well as drinking water
- Drinking different types of unhygienic water cause different diseases
- Most of the time of the women is spent arranging drinking water due to which the employment is affected
- The water table of the village is deep
- Erratic monsoon rains due to climate changes
- Saline and fluoride water problem – affected the lives of people (having problem of stone in kidney, dental and skeletal fluorosis), both in agricultural and drinking water front.

To address water scarcity challenges, a village-level water committee formed by dedicated women raised awareness about rainwater harvesting and drinking water protection. Through persistent meetings and petitions to the government, they secured resources to implement solutions. Their tireless efforts led to the construction of rainwater harvesting tanks, ponds with comfortable seating, and the recharge of existing wells, ensuring a reliable source of clean drinking water for the villagers. Furthermore, for agricultural needs, dams, check dams, and farm cisterns were built to capture rainwater. Additionally, irrigation wells were recharged, promoting sustainable water use for farming activities.

The campaign's success is evident; over 30,000 women from 400 villages remain actively involved in water management. They contribute to pump maintenance, reviving traditional water sources, protecting watersheds, recharging groundwater, harvesting rainwater from rooftop and more.

Climate change intensifies hardships, with cyclones, floods, heatwaves and cold waves disproportionately affecting the poor in developing countries. Poor women, facing existing social and economic disadvantages are especially vulnerable. These vulnerable populations lack access to the limited social safety nets and insurance available. This includes a vast segment of the informal workforce: small farmers, agricultural workers, fisherwomen, dairy farmers, vendors, construction workers, waste recyclers, and many others who bear the brunt of these increasingly frequent and devastating climate shocks.

For SEWA's 50th anniversary, its members made a ground-breaking decision; climate action would be the central theme for the next 50 years. "Building Clean Skies" became a core mission, driving a series of pioneering grassroots initiatives focused on climate resilience, adaptation and mitigation.

SEWA recognizes the knowledge gap farmers face regarding efficient irrigation methods. To address this, demonstrations were conducted on drip irrigation and solar pumps at various locations, inviting neighbouring farmers to learn. Additionally, SEWA helps finance the installation of drip irrigation systems, making them more accessible despite the initial cost. Farmers can then pay in instalments for a more manageable financial burden.

SEWA's training extends beyond irrigation. Across nine districts in Gujarat (Aravalli, Surendranagar, Chhota Udepur, Ahmedabad, Gandhinagar, Anand, Mahesana, Radhanpur and Kachchh), they train women farmers on various climate-smart practices. This includes raising awareness about solar technology like pumps, light traps and fencing. Additionally, they provide training on beneficial practices like mulching, organic farming and intercropping systems.

SEWA's *'Swachh Akash Campaign'* has demonstrably reduced water usage. Through 65 demonstrations held across Gujarat, Maharashtra, Rajasthan, and Uttar Pradesh, they've helped save an impressive 1.17 crore litres of water. This success has generated significant interest among farmers in setting up drip irrigation and solar pump systems. Recognizing the growing demand, SEWA offers financing options that make these systems more affordable. Farmers can pay in instalments, easing the initial financial burden. The benefits are twofold; conserving water and reducing energy consumption.

So let's all adopt climate-smart practices and build a future where; (i) our farmers and farms thrive by using water efficiently and adapting to changing weather patterns for ensuring the security and sustainability of our agricultural sector; (ii) our lives and livelihoods flourish attributed to sustainable practices that safeguard our food security, water resources and overall well-being; and (iii) our environment remains clean by reducing impact on the planet, and to create a healthier world for generations to come. Overall, our activities emphasize on the collective responsibility for a sustainable future and highlight the interconnected benefits of climate-smart practices.

Water Security for Agriculture—*Atmanirbhar* Farmers

Kanhaiya Choudhary[1]

Context

Water is a critical input into agriculture having a determining effect on the eventual yield. Without optimal watering, even the good seeds and other important inputs are not able to achieve their full potential. Adequate availability of water is important for animal husbandry as well. India accounts for about 17% of the world's population but only 4% of the world fresh water resources. India ranks 2nd world wide in farm output. India is the world's largest producer of fresh fruits and vegetables, milk, major spices, various crops such as jute, staples such as millets and castor oil seed. It is also the second largest producer of wheat and rice. Agriculture and allied sectors like forestry and fisheries accounted 13.7% of the GDP (Gross Domestic Production) in 2013 and employed 50% of the workforce. The irrigation infrastructure includes a network of canals from rivers, ground water, well based systems, tanks and other rain water harvesting products for agriculture activities. Today ground water system is the largest, covering – 160 million ha of cultivated land in India with 39 million ha irrigated by ground water, 22 million ha by irrigated canals and about two third of cultivation in India is still depending on monsoon. With 1544 m^3 per capita water availability, India is already a water-stressed country and is moving towards turning into water scarce. India is not a water rich country and is further challenged due to negative impact of climate change; enormous wastage owing partly to poor management and distorted water pricing policies.

Eighty percent of India's fresh water is used in agriculture. India receives an average of 4,000 billion cubic meters of precipitation every year. However, only 48% of it is used in India's surface and groundwater bodies. A dearth of storage procedure, lack of adequate infrastructure, inappropriate water management has created a situation where only 18–20% of the water is actually used. India's annual rainfall is around 1183

[1] CEO, N.M. Sadguru Water and Development Foundation

mm, out of which 75% is received in a short span of four months during monsoon (July to September). This result in run offs during monsoon and calls for irrigation investments for rest of the year. The population of India is likely to be 1.6 billion by 2050, resulting in increased demand for water, food and energy. This calls for infrastructure expansion and improved resource utilization. It is worth mentioning that climate change is also posing its challenges on agricultural productivity ranging from crop selection; time of cultivation, irrigation methods etc.

N.M. Sadguru's Work Area

N.M. Sadguru Water and Development Foundation is working in the tribal heart land of Gujarat, Rajasthan and Madhya Pradesh with its major presence in Dahod, Panchmahals, Banswara, Dungarpur & Jhalawar and Jhabua, Alirajpur, Barwani and Mandsaur districts. Water deficiency is a critical factor that limits growth and development and India ranks high on water stress. This issue is more in the tribal region of India where there is very limited access to water for irrigation.

N.M. Sadguru's project area have been among the most backward regions of our country. It is very well known for its abysmal poverty and all kind of problems resulting from such poverty. General scenario of the region is that Irrigation coverage is less than 20% and agriculture productivity is less than 50% compared to the state average productivity. The status of the project village is more vulnerable where irrigation ranges from as low as 1.7% to 20%. Average agriculture productivity is only 44% of the national average and much less as compared to district and state average.

N.M. Sadguru's Initiatives

The livelihood of people in N.M. Sadguru's project area is heavily reliant on agriculture. However, due to the lack of water security, agriculture alone cannot provide food security, leading to forced migration to urban areas for livelihood. Recognizing this challenge and noting that nearly all tribal households own land, the organization devised a plan for the region's most effective development through land and water resource management. The initial objectives were centred around developing water resources, planting boundaries of farmers' fields, and building community capacity for decentralized asset management to ensure food security and environmental enrichment.

With encouraging outcomes, the initiatives expanded to include changing cropping patterns to incorporate cash crops like vegetables, spices and fruits, as well as off-farm activities such as dairy, goat farming, and poultry to increase household income alongside ensuring food security. This approach has enabled farmers to sustainably support their livelihoods in their own villages, improving their quality of life. As a result, N.M. Sadguru's development journey for rural and tribal communities has transformed them from poverty to food security and ultimately to prosperity. This comprehensive model, encompassing water resources development, scientific agricultural practices and community-managed dairy, has evolved into a successful integrated model for rural livelihoods.

Water Security Initiatives

N.M. Sadguru has played a vital role in improving access to irrigation for small and marginal farmers by focusing on water conservation, optimal harvesting and judicious water use. The organization has primarily concentrated on two key areas: promoting appropriate and improved technologies in water conservation, harvesting and agriculture; and enhancing community ownership of the processes and assets involved. Despite the region's average annual rainfall of 800 mm, there is a significant water scarcity, primarily due to poor water management in the hilly terrain. Although there are networks of rivers and rivulets, runoff quickly drains out of the area.

To address this issue, N.M. Sadguru has implemented a strategy to optimize runoff harvesting in all rivers, ensuring that the entire river length has water storage capacity, thus providing water security to every village. Surface water harvesting has been achieved through the construction of a series of check dams along the entire length of rivers and rivulets, as well as the construction of ponds and tanks in strategic locations in the watershed area where rivers and major rivulets are not present. In total, more than 1250 structures have been built, creating surface storage for 1,68,000 million litres of water. This effort has significantly recharged the groundwater aquifer, leading to increased well yields and the availability of well water throughout the summer season.

Moreover, *in situ* moisture conservation has been achieved through various soil and water conservation measures in the watershed area, further enhancing groundwater recharge and ensuring sufficient soil moisture for rain-fed crops.

To improve irrigation facilities, especially in the hilly terrain, a large number of community lift irrigation schemes have been implemented.

Additionally, micro-irrigation systems (such as drip, sprinkler, laser spray and drum kit gravity drip) and solar mini-irrigation systems based on wells have been promoted according to the needs and affordability of the community. Overall, 1,55,000 acres of land belonging to approximately 1,20,000 households have been brought under irrigation. Watershed treatment has transformed 1,25,000 acres of land from low-yield to high-yield cropping patterns. One of the most significant aspects of this initiative has been the decentralized management of irrigation systems and water resources by the community. Farmers' collectives own, manage and maintain these systems at their own cost and effort.

Improved Agriculture

With assured access to irrigation, farmers receive training and capacity building on standard agricultural practices for cereals, pulses and horticulture. This has led to the adoption of high-value crops such as vegetables, flowers, fruits and spices. Currently, 2,52,000 farmers are engaged in vegetable and spice cultivation, 8,350 in floriculture and 53,200 in fruit crops. The yield of cereal crops has increased by 100 to 150%, providing not only food security but also a substantial income. Vegetable cultivation generates income in the range of Rs. 50,000 to Rs. 60,000 per 0.25 acres of land, while fruit crops yield Rs. 60,000 to Rs. 75,000 per 0.75 acres.

Farmers, under the leadership of farmers' federations, are also involved in seed production of maize, wheat, soybean, onion and bitter gourd. Hi-tech vegetable seedling nurseries have been promoted with entrepreneurial farmers. These initiatives aim to establish an ecosystem for improved agriculture, ensuring quality inputs are available timely and at an affordable price, which is crucial for the sustainability of any agricultural initiative.

Peripheral Plantation

Initially, plantation on the boundaries of land parcels was initiated with the purpose of creating assets for farming families. This activity later became a core focus, as plantation also enriches the environment, creating a better ecosystem for improving soil moisture conditions. The N.M. Sadguru Foundation promoted agroforestry and peripheral plantation, and today, 1.63 lakh farmers have planted 722 lakh trees, which is playing an important role in the overall ecosystem.

Animal Husbandry

The availability of water has enabled small and marginal farmers to engage in agriculture-based livelihoods and adopt dairy, goat rearing and poultry as part of their allied activities. Women's Self-Help Groups have taken the lead in this, and more than 25,000 households have adopted these activities, providing an additional income of Rs. 30,000 to Rs. 40,000.

Community Institutions

N.M. Sadguru has promoted a large number of primary farmers' collectives and apex federations to promote ownership and decentralized management. Community institutions are playing important roles in sustaining and scaling up these activities.

Primary Farmers' Collectives

Primary farmers' collectives include formal irrigation cooperatives, informal water user groups, self-help groups, milk cooperatives and farmer producer organizations (FPOs). There are 3020 primary farmers' collectives, mostly at the village level and centred around the initiatives.

Apex Farmers' Collectives- Federation

Apex farmers' collectives include irrigation federations, self-help group (SHG) federations and horticulture cooperatives. There are 12 apex federations at the taluka and district levels. These farmers' collectives are spearheading initiatives such as establishing plant nurseries for raising healthy seedlings on a large scale, seed production of various crops, supplying quality agricultural inputs, processing spices and millets and marketing cereals, vegetables, and fruits.

It can be concluded that water security has been the key to the success of all the development efforts in N.M. Sadguru's work area. The organization has prioritized water security right from the beginning, addressing a heartfelt need of the community and the region. Agriculture-based livelihoods have become remunerative and regenerative with water security.

References

[1] N.M. Sadguru Annual Report-2023.

[2] Climate Change Mitigation in India's Drylands by Dr. G. Agoramurthi.

[3] Water and Agriculture in India Background paper for the South Asia expert panel during the Global Forum for Food and Agriculture (GFFA) 2017 - Dr. Vibha Dhawan.

Water Security for Agriculture Sector

Kartik P. Dongre[1]

Groundwater is the backbone of India's agriculture and drinking water security, serving as a common-pool resource for millions of farmers across the country and often remaining the sole drinking water source for rural households. It also supports industrial water demand in many cases. Currently, more than 85 percent of India's rural domestic water requirements, 50 percent of its urban water requirements and over 60 percent of its irrigation requirements are being met from groundwater resources. During prolonged drought periods, groundwater becomes the only dependable source of water supply.

The increasing dependence on groundwater as a reliable water source has led to its large-scale and often indiscriminate development in various parts of the country. Many people still believe that groundwater is an unlimited resource and that more groundwater is available as one goes deeper. However, it is now recognized that groundwater is not an infinite resource and its availability in space and time is limited. In light of these facts, there is a pressing need for strategies aimed at the scientific and sustainable management of available groundwater resources in the country. Such strategies should address not only increasing groundwater availability but also Regulatory & Protection Measures and Public Awareness regarding various aspects of groundwater.

According to the Food and Agriculture Organization of the United Nations, 17% of food available to consumers worldwide is thrown away each year, wasting precious water used in its production. Wetlands, the most biologically diverse of all ecosystems, are disappearing three times faster than forests, with approximately 85% lost over the past 300 years. Global water demand for agriculture is expected to increase by 35% by 2050. Since 2000, flood-related disasters have increased by 134%, and the number and duration of droughts have increased by 29%.

Thus, a change in approach can alter the entire socio-economic scenario and this change can be achieved through:

- Shifting from Supply Side to Demand Side Management

---

[1] Scientist-D and Head of Office, CGWB, Ahmedabad.

- Implementing Artificial Recharge and Water Conservation Measures
- Emphasizing Participatory Management
- Encouraging Self-Regulation instead of Government Regulation.

Climate Change and Water Security for Agriculture Sector: Meetings Proceedings

Shrikant S. Kalamkar and Kirit Shelat[1]

Preamble

India is an agriculture economy where land and water are two key natural resources upon which farmers depend for their livelihoods and development. Farmers' development depends upon interactions of these and other resources, institutions, actions and policies and their ultimate outcomes. It would be naive to perceive that all rural poverty problems could be solved through improving the poor's access to water alone through development of irrigated area in rainfed conditions. However, though water is only a single element in the poverty equation, it plays a disproportionately powerful role through its wider impacts on such factors as food and other essential agricultural production. Water is one of the most critical inputs for agriculture and availability of adequate water for irrigation is a key factor in achieving higher productivity. Therefore, water security is importance for further growth of agriculture in general and food production in particular.

The world is facing many challenges; prominent among these are climate change, environmental degradation, sustainable development and food security for growing population. The threat of climate change is now considered as an established fact and talked as a global common problem. The Intergovernmental Panel on Climate Change (IPCC) in its fourth assessment report had observed that, '*the warming of the climate system is now unequivocal, as is now evident from observations of increases in global average air and ocean temperatures, widespread melting of snow and ice, and rising global sea levels*' (Solomon, *et al.*, 2007). About two billion people worldwide don't have access to safe drinking water today (SDG Report 2022), and roughly half of the world's population is experiencing severe water scarcity for at least part of the year (IPCC). These numbers are expected to increase, exacerbated by climate change and population

[1] Director Professor, Agro-Economic Research Centre, Sardar Patel University, Vallabh Vidyanagar & Former Secretary (IAS) Agriculture, Govt of Gujarat ad Chairman, NCCSD, Ahmedabad.

growth. Climate change is a real and undeniable threat to our entire civilization. The effects are already visible and will be catastrophic unless we act now. Climate change is exacerbating both water scarcity and water-related hazards (such as floods and droughts), as rising temperatures disrupt precipitation patterns and the entire water cycle.

Water and climate change are inextricably linked. Climate change affects the world's water in complex ways. From unpredictable rainfall patterns to shrinking ice sheets, rising sea levels, floods and droughts–most impacts of climate change come down to water. Through education, innovation and adherence to our climate commitments, we can make the necessary changes to protect the planet. These changes also provide huge opportunities to modernize our infrastructure which will create new jobs & promote greater prosperity across the globe.

Climate change will impact the extent and productivity of both irrigated and rainfed agriculture. Rising temperatures will translate into increased crop water demand, so will be demand of increasing Urbanities. Both the livelihoods of rural communities and the food security of a predominantly urban population are therefore at risk from water-related impacts linked primarily to climate variability. The rural poor, who are the most vulnerable, are likely to be disproportionately affected. Various adaptation measures that deal with climate variability and build upon improved land and water management practices have the potential to create resilience to climate change and to enhance water security. In order to meet the challenge of food security, there is a need to understand need of water security.

In context of having more in-depth research and deliberations on the issues mentioned above, Agro-Economic Research Centre of Sardar Patel University and National Council for Climate Change and Sustainable Development and Public Leadership, Ahmedabad have entered into an agreement for action research on climate change. Agreement focuses on academic collaboration to bridge gap between academic research, innovation and adoption and policy development through systematic research and extension for prevention of impact of climate change. Both the Institutions have agreed to work together on climate change and its impact on water cycle, water security, and that in turn on food security. "Climate Smart Agriculture Unit" is established at AERC, for training and capacity buildings of stakeholders to mitigate the impact of climate change and to popularise the adaptation strategies.

In the follow up of above targets, both THE institutions have jointly organised three national seminars cum interaction meets 'Climate Change and Water Security for Agriculture Sector' on June 3, 2023; August 21, 2023 and December 22, 2023. The deliberations were on Climate change and its impact on water cycle, water security, and that in turn on Food Security. The issues related were discussed at length and come out with the action plan. The proceedings of these three meets are summarised below:

1. *Climate Change & Water Security – June 3, 2023*

The first national seminar cum interaction meet was organised at Agro-Economic Research Centre, SPU, Vallabh Vidyanagar (Gujarat) on June 3, 2023. Shri Atul Shroff, Former Managing Director, SFT and Transpek Industry Ltd., Vadodara was the chief guest of the inaugural session and Dr. Kirit Shelat, IAS (Rtd), Former Principal Secretary, Government of Gujarat & Executive Chairman, NCCSD, Ahmedabad was the guest of honour. At the beginning of the national seminar, Dr. S.S. Kalamkar, Director of the AERC, Sardar Patel University, Vallabh Vidyanagar welcomed the gathering. Dr. Bhailalbhai Patel, Registrar of the University also welcomed the guest and participating experts and highlighted the uniqueness of AERC in organising such academic events on topics of national importance and proudly mentioned about AERC and Department of Economics as a star department of the University.

Shri Atul Shorff in his chief guest address mentioned that water and climate change are inextricably linked and climate change affects the world's water in complex ways. He highlighted that water scarcity in India is expected to worsen as the overall population is expected to increase to 1.6 billion by year 2050. Water scarcity is emerging as a formidable business risk for companies across the world and India too faces the same problem, magnified several times in the past few years. On one hand, we have acute shortage and on other-hand, we have situation of over precipitation and floods. Callous disposal of solid waste & untreated sewage is polluting the quality of ground water. He further mentioned that India generates huge quantity of sewage (around 60000 MLD) from every city, town and village. The large cities and towns have Centralized Sewage Systems, with underground pipes, pumping stations, and treatment plants. However, these systems are expensive to build and to operate, requiring uninterrupted power, skilled operators, and extensive maintenance. As a result, more than 70% of it is untreated and disposed causing pollution of water bodies. He suggested for treatment of waste water using a simple, decentralized, and environment friendly methods using minimal energy which is applicable to rural as well as urban areas.

Dr. Kirit Shelat mentioned the importance of water in sustainable livelihood and need to recognize water as a scarce resource and adopt less water activities. He suggested the need to meter water use and to begin charging for water with initially token charge. He emphasizes the need of water harvesting not only through dams and ponds, but by making compulsory recharge wells in all housing societies, infrastructure, industry and even on farms. Further, he advocated for water recycling including, waste water re-cycling which need to be made compulsory for all high rise buildings, private and public townships. Also to avoid wastage of surface water, canal water need to brought under piped water. There is a need to make compulsory use of micro irrigation systems, such as drip irrigation with subsidy. Sea-water ingress need to be prevented by protection walls. It needs to be monitored on year-to-year basis. India has enough water resources, but there is a need to have its responsible and efficient use by farmers, house-holders, industry etc. Also needed an efficient transport system and water storage structures which would help to continued sustainable development. He mentioned that with more demand for water in the arena of climate change, more energy is needed to pump underground water. Therefore, use of solar energy for desalination sea-water can enhance the water resources.

The eminent experts who have delivered their experts observations were Dr. S. V. Shah, Unit Head, Veterinary and Animal Science Research and Extension Unit, Kamdhenu University, Gandhinagar/Anand; Dr. K.N. Wadhwani, Research Scientist & Head, Livestock Research Station, Kamdhenu University, Gandhinagar; Dr. R. Sama, Former Director, Water and Sanitation Management Organisation, Ahmedabad; Dr. M.L. Gaur, Professor, Department of Agricultural Engineering, Anand Agricultural University, Anand; Dr. H.D. Rank, Professor & Head, Dept. of Soil & Water Conservation Engineering, CAET, Junagadh Agricultural University, Junagadh; Ms. Pratiti Priyadarshini, Head of Research and Learning, Foundation for Ecological Security, Vadod, Anand; Ms. Suchiradipta Bhattacharjee, International Water Management Institute, Anand; Ms. Laxmi Sharma, International Water Management Institute, Anand and Ms. Bhartiben Bhavsar, Rural Development Department, Self Employed Women's Association, Anand.

Dr. Kinjal Ahir, Honorable Deputy Director of Agro-Economic Research centre acted as Master of Ceremony and also proposed the vote of thanks at the end of the program.

Water Security for Agriculture Sector – August 21, 2023

The second national seminar cum interaction meet was organised on hybrid (virtual and offline) mode on August 21, 2023. Shri Raghavjibhai Hansrajbhai Patel, Honorable Minister of Agriculture, Animal Husbandry,

Cow-Breeding, Fisheries, Rural Housing and Rural Development, Government of Gujarat was the chief guest. At the beginning of discussion, Dr. Kalamkar, Director of AERC has presented in brief about the interaction meet and collaborative work started by AERC & NCCSD on Climate Change and Water security and related issues Dr. Niranjan Patel, Officiating Vice Chancellor of Sardar Patel University welcomed the Honorable Minister and delivered the opening remarks. Honorable Vice Chancellor emphasized that water is one of the most critical inputs for agriculture and availability of adequate water for irrigation is a key factor in achieving higher productivity. Therefore, water security is importance for further growth of agriculture in general and food production in particular. In order to secure vulnerable farmers from natural, social, and political barriers, it is required to provide them assure water for crop production and livestock.

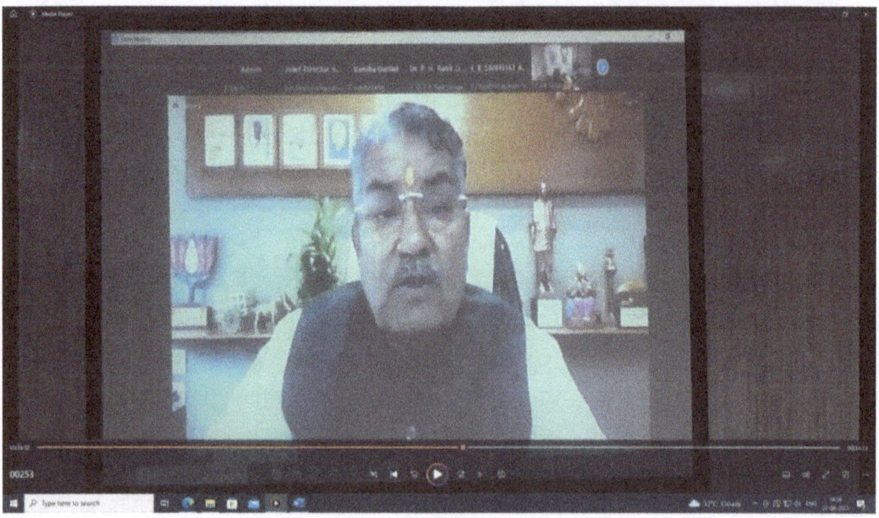

Honourable Minister Shri Raghavjibhai Hansrajbhai Patel highlighted the importance of water in increasing the crop productivity, crop production and thus income of the farmer. Minister sir suggested to find out the reasons as why farmers in Gujarat are not adopting micro irrigation on large scale. He also raised the issue of salinity, water logging and adverse climatic events and suggested that information should be provided to the farmers well in advance. He also mentioned about the schemes implemented by the Union Government for judicious use of water through micro irrigation technology such as drip and sprinkler. He suggested that during Amrit kal, farmers should be provided with information on water conservation and natural farming. He also wanted to think about to learn farmers to come out from the waterlogged condition due to heavy rain, farmers must get the proper information about the sudden cyclonic situation and how to overcome the situation and must learnt about the water conservation system. Lastly, he wanted recommendations from the interaction meet for further development in agriculture sector of the state.

Dr. Kirit Shelat delivered the address and highlighted the issues of climate change and rainfall pattern and thus judicious use of water. He emphasized on the loss of water during transport and thus advised to work on limit of water loses. He also said that agriculture is not only for the farmer, it is for all people and nation. Therefore, all universities, institutions, research centers should work on this topic to advocate water saving and use of water with technology.

Shri Jai Prakash Shivahare, IAS, Managing Director, Gujarat Green Revolution Company Ltd. highlighted major three issues, viz. GGRC model, crop diversification (connecting small farmers with horticulture along to agriculture) and behaviour change. He mentioned that government has increased fund for related water security schemes from Rs. 500 crore to Rs. 1500 crore. GGRC is introducing more and more farmers to the drip and sprinkler system for farming and it has also been successful. Drip and sprinkler system can be used to maximize the yield of water. Crop diversification also one factor affects to minimize water wastage in farming. Small and marginal farmers are engaged in agriculture only should be linked with horticulture. Farmers have to change their behaviour regarding new technology adoption so that they accept the technology. He also opined that other state has been adopting GGRC model and recently Madhya Pradesh adopted same.

Dr. A.R. Pathak, Former Vice Chancellor Junagadh and Navsari University noted that we have built dams and stored water although, there is a shortage of water in year by year so for that water has to be used sparingly and judiciously to everyone. When we talk about agriculture and irrigation, water pipeline should be started instead of open canal for irrigation. Instead of late sowing, farmers should sow according to the crops of the month and agricultural university's professors should coordinate with them and provide guidance farmers. Even agricultural university scientist and agriculture officers should work together on agriculture field. The area under drip and sprinkler has to be increased by encouraging them those who have irrigation cooperative societies in their villages. Farmers should also provide water to the crops as they need.

Dr. R. Sama, Former Director, Water and Sanitation Management Organisation (WASMO), Ahmedabad. He discussed about tribal area (Dediyapada, Narmada district) and use of micro irrigation under mini watersheds programme in this hilly area. In this area, there is server water shortages so he first made open wells in the village and collected rain water, then open well water used for drinking and irrigation. He give to new concept to the villagers first is Demand of water management and second is Water literacy.

Shri S.J. Solanki, Director (Agriculture), Directorate of Agriculture, Government of Gujarat, Gandhinagar highlighted that water scarcity is increasing day by day so water has to be delivered where there is no water and water has to be used sparingly where there is water. There is a need to encourage and increasing more areas under MIS. He said that fields have to be levelled with laser technology to so that increasing MIS coverage and where there are rainfed areas, water storage will have to be increased. To encourage natural farming and organic farming so that the quality of land will improve. If the quality of the soil improves, we will be able to store clean water recharge. To using jin technology and discovering new varieties of seeds so that we can use less irrigation water and get high productivity. To encourage and active irrigation cooperative societies in every village.

Shri Kartik P. Dongre, Scientist-D, Regional Director (I/c) Central Ground Water Board, West Central Region, Ahmedabad mentioned that only 4 percent of fresh water available in India and per capita water availability in India is very less. According to him, we consume water in two ways: supply side and demand side. In supply side, we should change mind approach about water and demand side, we should learn water management such as MIS system. We have saved 30 percent of water by using MIS technology. He opined that everyone should try for water recharge.

Dr. F.K. Modh, Joint Director of Horticulture, Mehsana said that out of 84 percent of water, 12% of water is used in industries and only 4 percent water is used households. Districts like Banaskantha, Kutch and Sabarkantha have a good prevalence of drip and sprinkler irrigation. In these areas farmers are also storage and adopt water recharge system.

Shri S.J. Solanki, Director, Agricultural Technology Management Agency (ATMA), Kheda mentioned that farmers should made efforts to conserve water by changing cropping patterns. Farmers should use canal water judiciously for irrigation. Appropriate measures should be taken by increasing the monitoring of the canal area where water is being wasted and measures should be taken to ensure that the marginal farmer can benefit from the canal water. He also emphasized on natural and organic farming so that we can save water.

Mr. R.H. Ladani, Joint Director of Horticulture, Rajkot opined that government should make water budget so that water can be saved. He said that, instead of increasing the area of sub-tropical crops, priority should be given to crops suitable for dry semi-arid areas. Instead of mono cropping method, mixed cropping method should be adopted. Waste land should be utilized such as cultivation of fruits and vegetables so that the waste land can be used (We should adopt model of Krip Home in Karnataka). We should adopt new technology such as eco block modules technology, jio modern technology so that we can storage water and recharge ground water. We should also adopt satellite/ET farming, crop cover technology and also adopt Israel farming technology.

Dr. M.L. Gaur, Professor, Department of Agricultural Engineering, B. A. College of Agriculture, Anand Agricultural University, Anand opined that we should make efforts of the farmers to make better farming by using this and adding more to them tradition knowledge in farming. He said we can change our food habits (For example, if more rice is used, more water is required to increase its production, but if we increase the use of coarse-grains, we will be able to get more nutritious grains with less water). We should develop rain storage ecosystem and irrigation ecosystem.

Dr. H.D. Rank, Professor & Head, Dept. of Soil & Water Conservation Engineering, CAET, Junagadh Agricultural University, Junagadh highlighted salinity related problems in Narmada-SSP command area and is management. He mentioned that 54 percent of India face high to extremely high-water stress condition. Water resource availability—Major and minor dams and their storage capacity to need to be restructured to meet demand for human population, livestock, season wise crops and industries areas. He said that Banaskantha, Sabarkantha, Gandhinagar, Mehsana, Patan, Arvalli districts are non-command area under Narmada canal, there is a salinity low in all these areas, so water saving can be increased. Central Gujarat and south Gujarat salinity level is high even Saurashtra region also affected salinity except Botad, Porbandar district. He suggested that, we should adopt more new technology in agriculture.

Smt. Bharatiben Bhavsar and Rehanaben Riyawala of Senior coordinators at SEWA foundation described working of SEWA foundation with special focus women empowerment in rural sector. They describe Soil Sahkar Campen, capacity building programme which was organized at village, block and district level, water resource management, deforestation and described also importance of solar pumps in agriculture.

Mr Mohan Sharma and Mr. Sachin Oza, Executive Director, Development Support Centre, Vadodara focused on promoting irrigation through water security. They elaborate twenty years journey of Development Support Centre for towards water security and participatory irrigation management. They emphasized water budgeting and water balancing, more crop per drop, ATAL Bhujal Yojna, solution of water demand. Their foundation was working in Lolasan village of Sabarkantha district and suggest very important findings through field experience like demand and water supply of available water resources in village and suggest valuable suggestion to manage water uses.

Shri Naveen Patidar, Chief Executive Officer, AKRSP-India on Soil Health and Water security highlighted about soil density, root penetration test, water infiltration test, organic matters in soil, water logging and soil stability test. They also discussed about soil test in laboratory with specific indicators. They conclude his lecture and very important message was told that good soil health improves water efficiency and security.

Dr Kaniyalal, CEO and Hitesh Shah, Sadguru foundation described working of foundation in watershed area treatment specially in hilly region of tribal area. They focus bunding in hilly area, converting waste land in agriculture land, ground level water management, efficient irrigation, cooperative lift irrigation, PM- KUSUM solar irrigation and many more community-based programme which was organized by their foundation.

Farmers from Aatapi Seva Foundation shared their experience on water management, animal husbandry, natural farming and horticulture development in respective area.

Shri V.N. Shendye, CSR project Coordinator-LOTE, Maharashtra elaborated his experience of different agriculture methods in Konkan area of Maharashtra.

Dr. Sonal Bhatt, Professor at Department of Economics, Sardar Patel University. Highlighted energy groundwater nexus in agriculture with special reference to water security. She told about Jyoti Gram Yojna and Soorya-Shakti Kisan Yojna for solar energy tariff rates in Gujarat.

The representatives of IWMI, Anand focused on SKY yojana, Dhundi village solar cooperative society and also described of PM-KUSUM yojana in India.

Overall Virtual Coordinator and Anchoring was done by Ms. Nisha Shah, Chief Executive Officer, NCCSD. Vote of thanks was delivered by the Dr. S.S. Kalamkar, Director, AERC. More than 100 participants had joined and participated in the discussion from various parts of our country and also those were traveling abroad.

3. December 22, 2023

The third national seminar cum interaction meet was organised on physical mode on December 22, 2023. Dr. V.K. Saraswat, Honorable Member NITI Ayog, Government of was the chief guest. At the beginning of discussion, Dr. Kalamkar, Director of AERC has presented summary of discussion held in last two interaction meets on Climate Change and Water security and related issues

Dr. Niranjan Patel, Officiating Vice Chancellor of Sardar Patel University welcomed the dignitaries and delivered the opening remarks. Honorable Vice Chancellor emphasized that water is one of the most critical inputs for agriculture and availability of adequate water for irrigation is a key factor in achieving higher productivity. Therefore, water security is importance for further growth of agriculture in general and food production in particular. He informed the gathering about 'Khedut Haat' started by the AERC on every Sunday to support farmers and help consumer to get quality food.

Dr. V.K. Saraswat, Honorable Member NITI Ayog, Government of India delivered chief guest address on this occasion. Dr. Saraswat mentioned that agriculture production is highly dependent in water and demand for same is increasing at very high rate. He emphasized on Green Hydrogen and water management as well as water stress condition in India. He cautioned about adverse impact of climate change on water availability and thus requirement of region wise plans, modeling and use of customized technology. He also raised the issue of water limited environments, declining yield, soil degradation, soil carbon storage, environmental quality and sustainability. He suggested that to make India as developed nation by 2047, we need to adopt technical strategies (blending water smart climate smart) for water secure agriculture in India.

Dr Meenesh Shah, Chairman, National Dairy Development Board addressed the gathering and highlighted that due to change in rainfall pattern, large areas of margin of desert have opened for taking up massive livestock program backed by Solar Energy distillation for drinking water. Dr Shah suggested that establishing dairy cooperatives, ensuring livestock health and nutrition, breed improvement programs, creating market linkage for local products, etc. would help improve livelihoods in the region. Solar distillation technology would play a pivotal role in sustainable water management. Adoption of sustainable practices, technological innovations and market-oriented strategies would be a key to the success of dairy production and better livelihoods.

Dr. Kirit Shelat highlighted the issues of water security in the context of increase in demand for water for agriculture and for domestic animals. He highlighted that though we have started using water saving technologies like drip and sprinkler, but its usage has limited coverage. We need to work on efficient canal water use through proper maintenance of canal, need to adopt leakage detection system and also monitor the transfer losses for judicious use of water. He emphasized on the loss of water during transport and thus advised to work on limit of water loses. He also said that agriculture is not only for the farmer, it is for all people and nation. Wastage of water at household level is also the important issue need to be addressed. Therefore, all universities, institutions, research centers should work on this topic to advocate water saving and use of water with technology, recycling of water and rainwater harvesting.

Dr. A.R. Pathak, Former Vice Chancellor, Junagadh Agricultural University, Junagadh highlighted that shortage of water resources is more severe due to the growth of population and human activities. Agriculture-food industry is the largest water-consuming industry, with more than 7090. Improving the resilience of water resource management is critical to reducing the risks in agriculture. Inefficient Water use Patterns also affect water demand and water supply in agri. & water management. Water is the primary resource of agriculture, and pressure to reduce consumption in line with conserving the environment and natural resources is essential. He suggested that in future, the food requirements of our ever-growing populations can be met from the domestic productions from the available water resources by enhancing the water productivities. The adoptions of MISs will not merely serve the purpose, as we are at the edge of the situation, where we have to think through the entire spectrum of the water resources issues. The water productivities from the agriculture sector can be enhanced through soil health management, assuring the water availabilities and critical stage wise crop water demands, deficit irrigation, mulching technology, irrigation water quality management, efficient operation and maintenance of MIS with new dimensions like automated AI and smart sensors-based irrigation and fertigation, pulse drip irrigation and aerated drip irrigation.

Dr. Sanjay Deshmukh, Professor of Life Sciences; Former Vice-Chancellor, University of Mumbai, Mumbai advocated enhancing water security for resilient agriculture in India and suggested pathways to sustainable livelihoods through water-resilient agriculture. viz. transformative shifts in agricultural landscape- an ecological perspective; high-value crop diversification strategies w.r.t. water security; adoption of agro-ecological practices to enhance water efficiency; harnessing technological innovations for efficient water use and market access; empowering agricultural communities through education and access to credit facilities; and building capacity to thrive in evolving dynamics of water security. The integration of efficient irrigation systems, rainwater harvesting, and soil moisture management techniques presents promising avenues for optimizing water resources. Moreover, the exploration of alternative crops that are well-suited to local agro-climatic conditions can diversify income streams for farmers. This diversification, coupled with access to markets for water-efficient produce, contributes to economic sustainability for farming communities. In the rain-fed regions of Andhra Pradesh, the adoption of precision agriculture techniques, coupled with the cultivation of climate-resilient crops like pulses and oilseeds, has transformed the agricultural landscape. This paradigm shift not only enhances water-use efficiency but also elevates farmers' incomes.

Dr. R. Krishnan, Director, Centre of Climate Change Research, Indian Institute of Tropical Meteorology, Pune mentioned that human activities affect all the major climate system components, with some responding over decades and others over centuries. He noted that global surface temperature will continue to increase until at least the mid-century under all emissions scenarios considered. Global warming of 1.50c and 20c will be exceed during the 21st century unless deep reductions in CO_2 and other greenhouse gas emission occur in the coming decades.

Dr. C.K. Timbadia, Vice Chancellor, Gujarat Natural Farming Science University, Halol, Dist. Panchmahal explored Natural Farming practice for enhancing water resilience and sustainability in agriculture. The research emphasizes the pivotal role of earthworms, in improving soil moisture retention, infiltration rate, and overall water-holding capacity. Additionally, the application of Jeevamrit in natural farming demonstrates significant increases in soil organic carbon, microbial diversity, and water-holding capacity. Combining zero tillage with cover crops proves effective in enhancing water availability and improving soil structure, while crop diversification emerges as a key strategy for water conservation. The study highlights the positive correlation between soil organic matter and moisture retention, underscoring the importance of these factors in mitigating the drought risks. By prioritizing these natural and regenerative approaches, agriculture can not only adapt to water-related challenges, but also contribute to the broader goals of sustainability and ecological balance.

Dr. Suresh Acharya, Former Director of Research, SDAU, SK Nagar & R&D Vertical, AGROCEL Industries Pvt. Ltd. Koday, Mandvi, Kachchh highlighted need of water resilient agriculture, use of technology in water budgeting, participation of community in decision making in distribution of water, use of precision agriculture technique and drought tolerant crops. He presented complexities of the SW-MEALS (Soil-Water-Microbes-Ecology-Agriculture-Livelihood Security) nexus in water-resilient agriculture requires a comprehensive strategy. Balancing the 'Demand Side' with the 'Supply Side,' weather conditions, and the natural resource base is crucial. The synergy between traditional methods and cutting-edge technologies is indispensable for a sustainable and resilient agricultural future. A participatory approach involving farmers and communities in local water budgeting, with incentives for water savings, is vital for promoting responsible water use. Innovative technologies, including smart irrigation systems, antitranspirants, anti-evaporation tools, and biocapsules, play a significant role in optimizing water management. Achieving water-resilient agriculture amid climate change requires a harmonious blend of traditional wisdom and innovative technologies, emphasizing the need for comprehensive strategies to enhance water resilience.

Dr. Ashwin Shroff in his article on 'Water Security and Novel Solutions for Agriculture' shared that Excel group has been actively engaged with Agriculture, farmers and rural communities. There are several initiatives, many for the first time in India and even globally e.g. Chemical pesticides, drip systems. Compost from organic wastes, sewage water treatment.

Rapporteurs for Sessions

June 3, 2023	*Rapporteurs*
Session I	: Shri Manish Makwana, AERC, SPU, VVN Dr. Thansigh Parihar, AERC, SPU, VVN
Session II	Dr. Kalpana Kapadia, AERC, SPU, VVN Ms. Aditee Sharma, AERC, SPU, VVN
Session III	Dr. Manishkant Ojha, AERC, SPU, VVN Dr. Jaydeepsinh K. Chauhan, AERC, SPU, VVN
August 21, 2023	
Session I	Dr. Hemant Sharma, Agricultural Economist Dr. Thansigh Parihar, AERC, SPU, VVN
Session II	Dr. Kalpana Kapadia, AERC, SPU, VVN Dr. Jaydeepsinh K. Chauhan, AERC, SPU, VVN
Session III	Dr. Manishkant Ojha, AERC, SPU, VVN & Shri Manish Makwna, AERC, SPU, VVN
December 22, 2023	
Session I	Dr. Hemant Sharma, Asst. Prof. AERC, SPU, VVN
Session II	Dr. Kalpana Kapadia, AERC, SPU, VVN Dr. Jaydeepsinh K. Chauhan, AERC, SPU, VVN
Session III	Shri Manish Makwna, AERC, SPU, VVN

Master of Ceremony

- Dr. Kinjal Ahir, Honorary Deputy Director of AERC and Head Department of Economics, SPU
- Ms. Nisha Shah, Chief Executive Officer, NCCSD
- Ms. Bhavini Aiya, NCCSD, Ahmedabad

Author Index